TEN-MINUTE EXERCISES

TEN-MINUTE EXERCISES

by

H. WEBB

To be used with
ARITHMETIC OF DAILY LIFE
By H. WEBB and J. C. HILL

TEACHERS' EDITION
WITH ANSWERS

CAMBRIDGE
AT THE UNIVERSITY PRESS
1938

CAMBRIDGE
UNIVERSITY PRESS

University Printing House, Cambridge CB2 8BS, United Kingdom

Cambridge University Press is part of the University of Cambridge.

It furthers the University's mission by disseminating knowledge in the pursuit of
education, learning and research at the highest international levels of excellence.

www.cambridge.org
Information on this title: www.cambridge.org/9781316601822

© Cambridge University Press 1938

First published 1938
First paperback edition 2015

A catalogue record for this publication is available from the British Library

ISBN 978-1-316-60182-2 Paperback

Cambridge University Press has no responsibility for the persistence or accuracy
of URLs for external or third-party internet websites referred to in this publication,
and does not guarantee that any content on such websites is, or will remain,
accurate or appropriate.

Introduction

The object of this series of books is to train the pupil—
   (a) in speed and accuracy in the fundamentals of arithmetic, and,
   (b) to avoid unnecessary written work.

Furthermore it is intended that the variety of easy problems should be used to interest the pupil in, and familiarise him with, new work which is introduced from time to time.

It will be seen that each one of the exercises consists of 10 questions. All 10 questions of each exercise should be answered by the normal Senior School pupil of eleven to thirteen years of age in 10 minutes. The answers only should be written down.

In each exercise the questions are graded. They commence very simply and increase slightly in difficulty as the exercise proceeds.

It will be found that the comparative simplicity of the exercises brings them within the scope of the backward child of a Senior School. In such cases *a time limit should not be imposed* and the exercises should be substituted for the longer type usually found in the Senior School Mathematics Syllabus. In this way mental fatigue will be avoided and much progress should be made through the stimulus of success which the pupil should obtain from working these short exercises.

Extracts from the Board of Education Handbook of Suggestions for the Teaching of Mathematics, 1937.

1. "Speed and accuracy count for most in the long run."

2. "Exercises should be so graded in difficulty that every child can enjoy the stimulus of success and of steady progress."

3. "Unnecessary written work should be avoided."

4. "In approaching new work the teacher should interest the pupil in a variety of easy problems involving such small numbers and such simple quantities that written work is not required."

5. "Few text-books contain enough exercises in Mental Arithmetic....If ample exercises are available one section of the class can be set down to 'mental work' by themselves."

Contents

(PART I)

(PART II)

(PART III)

PART ONE

Miscellaneous

Ex. 1A

1. Add together 6*d*.; 2*d*.; 9*d*.; 10*d*. Give your answer in shillings and pence.

2. 75 ÷ 5.

3. Take 1*s*. 8½*d*. from a florin.

4. 8 × 500.

5. At 3 for 2*d*. how many can I buy for 2*s*.?

6. In a class there were 27 present and 9 absent. How many were there in the class when all were present?

7. How many ounces are there in 1½ lb.?

8. What would be the cost of 5 handkerchiefs at 6½*d*. each?

9. How much shall I have to pay for 1½ lb. of beef at 1*s*. 6*d*. per lb.?

10. Tom buys a dozen oranges at 3 for 2*d*. What change will he receive from half-a-crown after paying for the oranges?

Ex. 1B

1. Add together 5*d*.; 4*d*.; 11*d*.; and 7*d*. Give your answer in shillings and pence.

2. 27 × 6.

3. Take 1*s*. 4½*d*. from half-a-crown.

4. How many fives are there in 195?

5. At 6 for 5*d*. how many can I have for 2*s*. 6*d*.?

6. In a class of 43 pupils 9 were absent. How many were present?

7. How many pints are there in 9½ gal.?

8. What would 4 books at 1*s*. 1½*d*. each cost?

9. What is the cost of 2½ lb. of butter at 1*s*. 3*d*. per lb.?

10. George buys half-a-dozen bananas at 2 for 2½*d*. How much change will he receive from a 2*s*. piece after paying for the bananas?

ANSWERS

Ex. 1 A

1. 2*s.* 3*d.*	2. 15.	3. 3½*d.*	4. 4000.	5. 36.
6. 36.	7. 24.	8. 2*s.* 8½*d.*	9. 2*s.* 3*d.*	10. 1*s.* 10*d.*

Ex. 1 B

1. 2*s.* 3*d.*	2. 162.	3. 1*s.* 1½*d.*	4. 39.	5. 36.
6. 34.	7. 76.	8. 4*s.* 6*d.*	9. 3*s.* 1½*d.*	10. 1*s.* 4½*d*

NOTES AND MEMORANDA

Addition, Subtraction and Multiplication

Below are extracts from two of the divisions of the Football League. Every team scores 2 points for a win and 1 for a draw.

Rewrite the tables and supply the missing figures.

Ex. 2A

Team	Played	Won	Drawn	Lost	Points
1. Chelsea	12	8	1	3	?
2. Wolves	12	7	2	?	16
3. Brentford	13	7	?	4	16
4. Charlton Ath.	12	?	5	2	15
5. Sunderland	?	7	1	4	15
6. Arsenal	12	?	4	3	14
7. Preston N. E.	12	5	?	3	14
8. Leeds Utd.	12	5	4	?	14
9. Stoke City	12	5	3	4	?
10. Bolton Wand.	?	5	3	4	13

Ex. 2B

Team	Played	Won	Drawn	Lost	Points
1. Coventry City	?	7	5	0	19
2. Sheffield Utd.	13	?	2	3	18
3. Aston Villa	12	8	?	3	17
4. Chesterfield	12	7	2	?	16
5. West Ham Utd.	12	5	5	2	?
6. Bradford	12	5	5	?	15
7. Norwich City	12	6	?	4	14
8. Burnley	13	?	4	4	14
9. Blackburn Rov.	?	4	5	3	13
10. Manchester Utd.	12	?	2	5	12

ANSWERS

Ex. 2A

1. 17.	2. 3.	3. 2.	4. 5.	5. 12.
6. 5.	7. 4.	8. 3.	9. 13.	10. 12.

Ex. 2B

1. 12.	2. 8.	3. 1.	4. 3.	5. 15.
6. 2.	7. 2.	8. 5.	9. 12.	10. 5.

NOTES AND MEMORANDA

Addition and Subtraction

1. Add 453
      9
     64
   —

2. Add £2. 17s. 6¾d.
      14s. 5½d.

3. Add 1 ton 16 cwt. 3 qr.
    2 ,, 18 ,, 3 ,,

4. Add 5 yd. 2 ft. 9 in.
    3 ,, 2 ,, 7 ,,

5. Take nine from twenty-six.

6. £1 − 11s. 6d.

7. From 2 yd. of ribbon I cut a piece 2 ft. 6 in. in length. What length have I left?

8. Take 28 pt. from 4 gal. 3 qt.

9. $36 - 4 - 5 - 7$.

10. John had 2s. 11d. in his money-box. To this he puts 6d. received from his mother, and 1s. 8d. that his uncle had given him. How much had he altogether?

1. Add 536
      8
     55
   —

2. Add £3. 16s. 5½d.
      12s. 7¼d.

3. Add 5 gal. 3 qt. 1 pt.
    4 ,, 2 ,, 1 ,,

4. Add 3 ton 14 cwt. 3 qr.
    1 ,, 19 ,, 2 ,,

5. From thirty-one take eight.

6. 10s. − 3s. 9d.

7. From a joint of beef weighing 5 lb. 4 oz. I cut a piece weighing 1½ lb. What was the weight of the remainder of the joint?

8. From 5½ gal. take 36 pt.

9. $48 - 9 - 5 - 4$.

10. George had 1s. 8d. and his father gave him 6d. If he spent 10d. of this money how much had he left?

ANSWERS

Ex. 3 A

1. 526.

2. £3. 12s. 0¼d.

3. 4 tons 15 cwt. 2 qr.

4. 9 yd. 2 ft. 4 in.

5. 17.

6. 8s. 6d.

7. 1 yd. 0 ft. 6 in.

8. 1 gal. 1 qt. or 10 pt.

9. 20.

10. 5s. 1d.

Ex. 3 B

1. 599.

2. £4. 9s. 0¾d.

3. 10 gal. 2 qt. 0 pt.

4. 5 tons 14 cwt. 1 qr.

5. 23.

6. 6s. 3d.

7. 3 lb. 12 oz.

8. 1 gal.

9. 30.

10. 1s. 4d.

NOTES AND MEMORANDA

Multiplication

Ex. 4

A. 1. 14×5.

    2. 47×9.

    3. 19×20.

    4. 2 ft. 3 in. $\times 6$.

    5. $1s.\ 8\frac{1}{2}d. \times 3$.

    6. 31×700.

    7. 2 tons 7 cwt. 2 qr. $\times 4$.

    8. £4. $5s.\ 2\frac{1}{2}d. \times 7$.

    9. How many ounces are there in 11 lb.?

   10. 1 gal. 3 qt. 1 pt. $\times 11$.

B. 1. 13×7.

    2. 39×8.

    3. 17×20.

    4. 2 lb. 6 oz. $\times 4$.

    5. $1s.\ 4\frac{1}{2}d. \times 4$.

    6. 41×500.

    7. 1 ton 12 cwt. 3 qr. $\times 3$.

    8. £3. $8s.\ 4\frac{1}{2}d. \times 8$.

    9. How many inches are there in 4 yd. 2 ft.?

   10. 2 gal. 2 qt. 1 pt. $\times 11$.

Division

Ex. 5

A. 1. $91 \div 7$.

    2. $288 \div 12$.

    3. $8500 \div 100$.

    4. 5 lb. 4 oz. $\div 3$.

    5. $4s.\ 1\frac{1}{2}d. \div 9$.

    6. $8250 \div 50$.

    7. £2. $6s.\ 8d. \div 4$.

    8. 4 yd. 2 ft. 9 in. $\div 6$.

    9. 6 gal. 3 qt. 1 pt. $\div 11$.

   10. I have to divide 17 lb. 3 oz. of beef between 5 people. What weight of beef will each receive?

B. 1. $85 \div 5$.

    2. $308 \div 11$.

    3. $9100 \div 100$.

    4. 6 ft. 8 in. $\div 4$.

    5. $4s.\ 11\frac{1}{2}d. \div 7$.

    6. $7320 \div 60$.

    7. £2. $3s.\ 9d. \div 3$.

    8. 3 yd. 2 ft. 8 in. $\div 8$.

    9. 9 gal. 1 qt. 1 pt. $\div 10$.

   10. I have to divide 20 lb. 4 oz. of beef between 6 people. What weight of beef will each receive?

ANSWERS

Ex. 4 A

1. 70.	2. 423.	3. 380.	4. 4 yd. 1 ft. 6 in.
5. 5s. 1½d.	6. 21,700.		7. 9 tons 10 cwt. 0 qr.
8. £29. 16s. 5½d.	9. 176.		10. 20 gal. 2 qt. 1 pt.

Ex. 4 B

1. 91.	2. 312.	3. 340.	4. 9 lb. 8 oz.
5. 5s. 6d.	6. 20,500.		7. 4 tons 18 cwt. 1 qr.
8. £27. 7s. 0d.	9. 168.		10. 28 gal. 3 qt. 1 pt.

Ex. 5 A

1. 13.	2. 24.	3. 85.	4. 1 lb. 12 oz.	5. 5½d.
6. 165.	7. 11s. 8d.	8. 2 ft. 5½ in.	9. $4\frac{7}{11}$ pt.	10. 3 lb. 7 oz.

Ex. 5 B

1. 17.	2. 28.	3. 91.	4. 1 ft. 8 in.	5. 8½d.
6. 122.	7. 14s. 7d.	8. 1 ft. 5½ in.	9. 3 qt. 1½ pt.	10. 3 lb. 6 oz.

NOTES AND MEMORANDA

Reduction

Ex. 6 A

1. Change 27 pence to shillings and pence.
2.   ,,   51 inches to yd., ft. and in.
3.   ,,   45 pints to gal., qt. and pt.
4.   ,,   35 ounces to lb. and oz.
5.   ,,   77 half-crowns to £ $s.$ $d.$
6.   ,,   5$s.$ 7½$d.$ to three-halfpences.
7.   ,,   £3. 10$s.$ 0$d.$ to three-and-fourpences.
8.   ,,   2 tons 7 cwt. to quarters.
9. How many seconds are there in 1 hr. 40 min.?
10. Add together 37 farthings, 26 pence, 5 sixpences, and 3 shillings. Give your answer in shillings and pence.

Ex. 6 B

1. Change 23 pence to shillings and pence.
2.   ,,   63 inches to yd., ft. and in.
3.   ,,   37 pints to gal., qt. and pt.
4.   ,,   29 ounces to lb. and oz.
5.   ,,   65 florins to £ $s.$ $d.$
6.   ,,   7$s.$ 4½$d.$ to three-halfpences.
7.   ,,   26 six-and-eightpences to £ $s.$ $d.$
8.   ,,   3 tons 5 cwt. to quarters.
9. How many yards are there in five miles?
10. Add together 27 halfpence, 31 pence, 7 sixpences, and 2 shillings. Give your answer in shillings and pence.

ANSWERS

Ex. 6 A

1. 2*s*. 3*d*.

2. 1 yd. 1 ft. 3 in.

3. 5 gal. 2 qt. 1 pt.

4. 2 lb. 3 oz.

5. £9. 12*s*. 6*d*.

6. 45.

7. 21.

8. 188.

9. 6000.

10. 8*s*. 5¼*d*.

Ex. 6 B

1. 1*s*. 11*d*.

2. 1 yd. 2 ft. 3 in.

3. 4 gal. 2 qt. 1 pt.

4. 1 lb. 13 oz.

5. £6. 10*s*. 0*d*.

6. 59.

7. £8. 13*s*. 4*d*.

8. 260.

9. 8800.

10. 9*s*. 2½*d*.

NOTES AND MEMORANDA

The Four Rules and Reduction

Ex. 7A

1. $17 + 6 + 9 + 5$.

2. How much is left after taking 4s. $7\frac{1}{2}d$. from 3 florins?

3. 2 ft. 11 in. ÷ 5.

4. How many pence are there in 8s. 5d.?

5. 23×700.

6. What is the cost of 4 articles at 3s. $1\frac{1}{2}d$. each?

7. How much shall I have to pay for 37 three-halfpenny stamps?

8. How many gallons and quarts are there in 76 pints?

9. Multiply half-a-gross by half-a-score.

10. 1 ton 7 cwt. 2 qr. of potatoes was equally divided between 11 families. What weight of potatoes did each family receive?

Ex. 7B

1. $16 + 7 + 8 + 4$.

2. How much is left after taking 6s. $1\frac{1}{2}d$. from 3 half-crowns?

3. 3 ft. 6 in. ÷ 7.

4. How many pence are there in 6s. 10d.?

5. 600×17.

6. What is the cost of 6 articles at 2s. $1\frac{1}{2}d$. each?

7. How many three-halfpenny stamps can I buy with 5s. $4\frac{1}{2}d$.?

8. How many yards and feet are there in 84 inches?

9. Divide one gross by half-a-dozen.

10. 2 yd. 2 ft. 4 in. of ribbon was equally divided between 5 girls. What length of ribbon did each receive?

ANSWERS

Ex. 7 A

1. 37.	2. 1*s.* 4½*d.*	3. 7 in.	4. 101.	5. 16,100.
6. 12*s.* 6*d.*	7. 4*s.* 7½*d.*	8. 9 gal. 2 qt.	9. 720.	10. 2 cwt. 2 qr.

Ex. 7 B

1. 35.	2. 1*s.* 4½*d.*	3. 6 in.	4. 82.	5. 10,200.
6. 12*s.* 9*d.*	7. 43.	8. 2 yd. 1 ft.	9. 24.	10. 1 ft. 8 in.

NOTES AND MEMORANDA

Short Methods in Multiplication and Division

Ex. 8 A.
1. 56×1000.
2. $9000 \div 100$.
3. 19×20.
4. $360 \div 40$.
5. 24×25.
6. $175 \div 25$.
7. 16×125.
8. $2500 \div 125$.
9. 16×101.
10. 14×99.

B.
1. 63×1000.
2. $6000 \div 100$.
3. 13×30.
4. $280 \div 20$.
5. 28×25.
6. $125 \div 25$.
7. 24×125.
8. $1500 \div 125$.
9. 17×101.
10. 15×99.

Costs of Dozens

Ex. 9

A. Find the cost of 1 dozen at:
1. $2d$. each.
2. $8d$. each.
3. $4\frac{1}{2}d$. each.
4. $6\frac{1}{4}d$. each.
5. $7\frac{3}{4}d$. each.

Find the cost of 3 dozen at:
6. $10\frac{1}{2}d$. each.
7. $1s$. $2\frac{3}{4}d$. each.

Find the cost of 1 at:
8. $3s$. $6d$. per doz.
9. $10s$. $9d$. per doz.
10. $14s$. $3d$. per doz.

B. Find the cost of 1 dozen at:
1. $3d$. each.
2. $10d$. each.
3. $3\frac{1}{2}d$. each.
4. $5\frac{1}{4}d$. each.
5. $9\frac{3}{4}d$. each.

Find the cost of 3 dozen at:
6. $11\frac{1}{2}d$. each.
7. $1s$. $4\frac{1}{4}d$. each.

Find the cost of 1 at:
8. $6s$. $6d$. per doz.
9. $8s$. $3d$. per doz.
10. $15s$. $9d$. per doz.

ANSWERS

Ex. 8 A

1. 56,000.	2. 90.	3. 380.	4. 9.	5. 600.
6. 7.	7. 2000.	8. 20.	9. 1616.	10. 1386.

Ex. 8 B

1. 63,000.	2. 60.	3. 390.	4. 14.	5. 700.
6. 5.	7. 3000.	8. 12.	9. 1717.	10. 1485.

Ex. 9 A

1. 2s.	2. 8s.	3. 4s. 6d.	4. 6s. 3d.	5. 7s. 9d.
6. £1. 11s. 6d.	7. £2. 4s. 3d.	8. 3½d.	9. 10¼d.	10. 1s. 2¼d.

Ex. 9 B

1. 3s.	2. 10s.	3. 3s. 6d.	4. 5s. 3d.	5. 9s. 9d.
6. £1. 14s. 6d.	7. £2. 8s. 9d.	8. 6½d.	9. 8¼d.	10. 1s. 3¼d.

NOTES AND MEMORANDA

Costs of Articles

Ex. 10 A. Find the cost of:

1. 7 articles at 4d. each.
2. 19 ,, at 1½d. each.
3. 27 ,, at 2s. each.
4. 31 ,, at 5s. each.
5. 16 ,, at 6s. 8d. each.
6. 42 ,, at 1s. 3d. each.
7. 240 ,, at 7½d. each.
8. 480 ,, at 3¼d. each.
9. 960 ,, at 5¾d. each.
10. 483 ,, at 9¼d. each.

B. Find the cost of:

1. 9 articles at 3d. each.
2. 21 ,, at 2d. each.
3. 33 ,, at 2s. 6d. each.
4. 29 ,, at 4s. each.
5. 19 ,, at 3s. 4d. each.
6. 38 ,, at 1s. 8d. each.
7. 240 ,, at 5½d. each.
8. 480 ,, at 6¼d. each.
9. 960 ,, at 4¾d. each.
10. 242 ,, at 8½d. each.

Costs of Weights

Ex. 11

A. What is the cost per lb. at:

1. ½d. per oz.?
2. 2d. per oz.?
3. 3d. per oz.?

At 1s. 4d. per lb. find:

4. The cost of 6 oz.
5. The cost of 1 lb. 5 oz.
6. The cost of 3 lb. 11 oz.

At 2s. per lb. find:

7. The cost of 4 oz.
8. The cost of 9 oz.
9. The cost of 1 lb. 9 oz.
10. The cost of 3 lb. 12 oz.

B. What is the cost per lb. at:

1. 1d. per oz.?
2. 1½d. per oz.?
3. 2½d. per oz.?

At 8d. per lb. find:

4. The cost of 10 oz.
5. The cost of 1 lb. 3 oz.
6. The cost of 2 lb. 13 oz.

At 2s. 8d. per lb. find:

7. The cost of 4 oz.
8. The cost of 7 oz.
9. The cost of 1 lb. 9 oz.
10. The cost of 2 lb. 14 oz.

ANSWERS

Ex. 10 A

1. 2s. 4d. 2. 2s. 4½d. 3. £2. 14s. 0d. 4. £7. 15s. 0d.
5. £5. 6s. 8d. 6. £2. 12s. 6d. 7. £7. 10s. 0d.
8. £6. 10s. 0d. 9. £23. 0s. 0d. 10. £19. 12s. 5½d.

Ex. 10 B

1. 2s. 3d. 2. 3s. 6d. 3. £4. 2s. 6d. 4. £5. 16s. 0d.
5. £3. 3s. 4d. 6. £3. 3s. 4d. 7. £5. 10s. 0d.
8. £12. 10s. 0d. 9. £19. 0s. 0d. 10. £8. 11s. 5d.

Ex. 11 A

1. 8d. 2. 2s. 8d. 3. 4s. 4. 6d. 5. 1s. 9d.
6. 4s. 11d. 7. 6d. 8. 1s. 1½d. 9. 3s. 1½d. 10. 7s. 6d.

Ex. 11 B

1. 1s. 4d. 2. 2s. 3. 3s. 4d. 4. 5d. 5. 9½d.
6. 1s. 10½d. 7. 8d. 8. 1s. 2d. 9. 4s. 2d. 10. 7s. 8d.

NOTES AND MEMORANDA

The Costs of Scores

Ex. 12

A. Find the cost of a score at:
1. 2s. each.
2. 16s. each.
3. 5s. 6d. each.
4. 7s. 3d. each.
5. 6s. 9d. each.

Find the cost of 1 at:
6. £4 per score.
7. £11. 10s. per score.
8. £23. 15s. per score.
9. £31. 5s. per score.
10. £16. 10s. per 3 score.

B. Find the cost of a score at:
1. 3s. each.
2. 15s. each.
3. 4s. 6d. each.
4. 8s. 3d. each.
5. 9s. 9d. each.

Find the cost of 1 at:
6. £7 per score.
7. £9. 10s. per score.
8. £25. 5s. per score.
9. £30. 15s. per score.
10. £13. 10s. per 3 score.

The Costs of Grosses

Ex. 13

A. Find the cost of 1 gross at:
1. 2d. each.
2. 7d. each.
3. 4½d. each.
4. 3¼d. each.
5. 6¾d. each.

Find the cost of 1 at:
6. £3. 12s. per gross.
7. £6 per gross.
8. £2. 5s. per gross.
9. £4. 7s. per gross.
10. £1. 10s. per gross.

B. Find the cost of 1 gross at:
1. 3d. each.
2. 6d. each.
3. 5½d. each.
4. 4¾d. each.
5. 7¼d. each.

Find the cost of 1 at:
6. £2. 8s. per gross.
7. £9 per gross.
8. £3. 3s. per gross.
9. £4. 13s. per gross.
10. £4. 10s. per gross.

ANSWERS

Ex. 12A

1. £2. 0s. 0d.
2. £16. 0s. 0d.
3. £5. 10s. 0d.
4. £7. 5s. 0d.
5. £6. 15s. 0d.
6. 4s.
7. 11s. 6d.
8. £1. 3s. 9d.
9. £1. 11s. 3d.
10. 5s. 6d.

Ex. 12B

1. £3. 0s. 0d.
2. £15. 0s. 0d.
3. £4. 10s. 0d.
4. £8. 5s. 0d.
5. £9. 15s. 0d.
6. 7s.
7. 9s. 6d.
8. £1. 5s. 3d.
9. £1. 10s. 9d.
10. 4s. 6d.

Ex. 13A

1. £1. 4s. 0d.
2. £4. 4s. 0d.
3. £2. 14s. 0d.
4. £1. 19s. 0d.
5. £4. 1s. 0d.
6. 6d.
7. 10d.
8. $3\frac{3}{4}d$.
9. $7\frac{1}{4}d$.
10. $2\frac{1}{2}d$.

Ex. 13B

1. £1. 16s. 0d.
2. £3. 12s. 0d.
3. £3. 6s. 0d.
4. £2. 17s. 0d.
5. £4. 7s. 0d.
6. 4d.
7. 1s. 3d.
8. $5\frac{1}{4}d$.
9. $7\frac{1}{4}d$.
10. $7\frac{1}{2}d$.

NOTES AND MEMORANDA

Miscellaneous Exercises

Ex. 14A

1. Add together 15, 7, 8 and 9.
2. From ten shillings take 3s. 8½d.
3. 4s. 4½d. ÷ 5.
4. 57 × 400.
5. Find the total cost of 2 doz. at 4 for 3d. and 8 at 1s. 6d. per doz.
6. What is the cost of 2 doz. articles at 3¼d. each?
7. 48 × 25.
8. Share 4s. between Tom and Mary giving Tom 8d. more than Mary.
9. A boy is 1 ft. 10½ in. shorter than his father who is 5 ft. 9¼ in. tall. How tall is the boy?
10. The distance between telegraph poles is 32½ yd. What is the distance between the first and fourth poles?

Miscellaneous Exercises (*continued*)

Ex. 14B

1. Add together 17, 7, 6 and 8.
2. From eight shillings take 5s. 3½d.
3. 2s. 7½d. × 8.
4. 350 ÷ 25.
5. Find the cost of 47 articles at 4s. each.
6. What will be the cost of a score of articles at 2s. 9d. each?
7. 64 × 125.
8. Share 6s. between Harry and Joan so that Harry has 3 times more than Joan.
9. How much shall I have to pay for 1 gross of articles costing 6½d. each?
10. Change 2645 yd. to miles and yards.

ANSWERS

NOTES AND MEMORANDA

Bills

Ex. 15

A. Find the cost of the following:

1. 16 articles at $1\frac{1}{2}d$. each.
2. 7 ,, at 2d. each.
3. 12 ,, at $7\frac{3}{4}d$. each.
4. 8 ,, at $5\frac{1}{4}d$. each.
5. 18 ,, at 1s. 5d. per doz.
6. 6 ,, at 3s. $4\frac{1}{2}d$. each.
7. $2\frac{1}{2}$ lb. at 1s. 7d. per lb.
8. 24 articles at 6 for 5d.
9. 14 oz. at 2s. 8d. per lb.
10. Find the total cost of the items above.

B. Find the cost of the following:

1. 24 articles at $\frac{1}{2}d$. each.
2. 8 ,, at 3d. each.
3. 12 ,, at $9\frac{1}{4}d$. each.
4. 9 ,, at $3\frac{3}{4}d$. each.
5. 30 ,, at 1s. 1d. per doz.
6. 5 ,, at 2s. $7\frac{1}{2}d$. each.
7. $3\frac{1}{2}$ lb. at 1s. 3d. per lb.
8. 42 articles at 7 for 6d.
9. 10 oz. at 2s. per lb.
10. Find the total cost of the items above.

Change

Ex. 16

A. What amount of change is received in the following cases?

	Money proffered	Amount of bill
1.	1s.	5d.
2.	1s.	$8\frac{1}{2}d$.
3.	2s.	1s. $2\frac{1}{2}d$.
4.	4s.	2s. 9d.
5.	2 half-crowns.	3s. 7d.
6.	7s.	5s. $2\frac{1}{2}d$.
7.	10s note.	7s. $1\frac{1}{2}d$.
8.	12s. 6d.	11s. 8d.
9.	£1.	15s. $9\frac{1}{2}d$.
10.	30s.	£1. 2s. 10d.

B. What amount of change is received in the following cases?

	Money proffered	Amount of bill
1.	1s.	7d.
2.	1s.	$4\frac{1}{2}d$.
3.	2s. 6d.	1s. $7\frac{1}{2}d$.
4.	3s.	1s. 11d.
5.	2 florins.	2s. 5d.
6.	8s.	6s. $3\frac{1}{2}d$.
7.	10s. note.	8s. $2\frac{1}{2}d$.
8.	15s.	12s. 7d.
9.	£1.	16s. $4\frac{1}{2}d$.
10.	30s.	£1. 5s. 8d.

ANSWERS

Ex. 15 A

1. 2*s*. 2. 1*s*. 2*d*. 3. 7*s*. 9*d*. 4. 3*s*. 6*d*. 5. 2*s*. 1¼*d*.

6. £1. 0*s*. 3*d*. 7. 3*s*. 11¼*d*. 8. 1*s*. 8*d*. 9. 2*s*. 4*d*. 10. £2. 4*s*. 9*d*.

Ex. 15 B

1. 1*s*. 2. 2*s*. 3. 9*s*. 3*d*. 4. 2*s*. 9¾*d*. 5. 2*s*. 8¼*d*.

6. 13*s*. 1½*d*. 7. 4*s*. 4½*d*. 8. 3*s*. 9. 1*s*. 3*d*. 10. £1. 19*s*. 6¼*d*.

Ex. 16 A

1. 7*d*. 2. 3½*d*. 3. 9½*d*. 4. 1*s*. 3*d*. 5. 1*s*. 5*d*.

6. 1*s*. 9½*d*. 7. 2*s*. 10½*d*. 8. 10*d*. 9. 4*s*. 2½*d*. 10. 7*s*. 2*d*.

Ex. 16 B

1. 5*d*. 2. 7½*d*. 3. 10½*d*. 4. 1*s*. 1*d*. 5. 1*s*. 7*d*.

6. 1*s*. 8¼*d*. 7. 1*s*. 9½*d*. 8. 2*s*. 5*d*. 9. 3*s*. 7½*d*. 10. 4*s*. 4*d*.

NOTES AND MEMORANDA

Perimeters of Rectangles

Ex. 17

A. Find the perimeters (distance all the way round) of rectangles having the following dimensions:

	Length	Breadth
1.	4 in.	4 in.
2.	9 in.	5 in.
3.	12 in.	7 in.
4.	1 ft. 2 in.	8 in.
5.	$10\frac{1}{4}$ in.	$4\frac{1}{2}$ in.
6.	1 ft. 10 in.	1 ft. 10 in.
7.	2 ft. $4\frac{1}{2}$ in.	1 ft. 7 in.
8.	$\frac{1}{4}$ ml.	200 yd.
9.	1 ft. $6\frac{3}{4}$ in.	$11\frac{1}{2}$ in.
10.	$2\frac{1}{2}$ yd.	2 ft. 3 in.

B. Find the perimeters (distance all the way round) of rectangles having the following dimensions:

	Length	Breadth
1.	3 in.	3 in.
2.	8 in.	7 in.
3.	11 in.	9 in.
4.	1 ft. 4 in.	6 in.
5.	$10\frac{3}{4}$ in.	$5\frac{1}{2}$ in.
6.	2 ft. 9 in.	2 ft. 9 in.
7.	2 ft. $7\frac{1}{2}$ in.	1 ft. 4 in.
8.	$\frac{1}{2}$ ml.	100 yd.
9.	1 ft. $8\frac{1}{4}$ in.	$10\frac{3}{4}$ in.
10.	$2\frac{1}{2}$ yd.	1 ft. 9 in.

Areas of Rectangles

Ex. 18

A. Find the areas of rectangles having the following dimensions:

N.B. In each case give your answer in square inches

	Length	Breadth
1.	7 in.	7 in.
2.	8 in.	5 in.
3.	11 in.	9 in.
4.	1 ft. 5 in.	6 in.
5.	$10\frac{1}{2}$ in.	8 in.
6.	$6\frac{3}{4}$ in.	5 in.
7.	2 ft. $3\frac{1}{2}$ in.	11 in.
8.	1 ft. 3 in.	$9\frac{1}{2}$ in.
9.	$9\frac{1}{2}$ in.	$5\frac{1}{2}$ in.
10.	1 ft. $7\frac{1}{4}$ in.	7 in.

B. Find the areas of rectangles having the following dimensions:

N.B. In each case give your answer in square inches

	Length	Breadth
1.	6 in.	6 in.
2.	9 in.	7 in.
3.	12 in.	8 in.
4.	1 ft. 3 in.	5 in.
5.	$11\frac{1}{2}$ in.	6 in.
6.	$7\frac{1}{4}$ in.	5 in.
7.	2 ft. $7\frac{1}{2}$ in.	10 in.
8.	1 ft. 5 in.	$7\frac{1}{2}$ in.
9.	$8\frac{1}{2}$ in.	$6\frac{1}{2}$ in.
10.	1 ft. $6\frac{3}{4}$ in.	9 in.

ANSWERS

Ex. 17A

1. 1 ft. 4 in.   2. 2 ft. 4 in.   3. 3 ft. 2 in.   4. 3 ft. 8 in.
5. 2 ft. 5½ in.   6. 2 yd. 1 ft. 4 in.   7. 2 yd. 1 ft. 11 in.
8. 1280 yd.   9. 1 yd. 2 ft. 0½ in.   10. 6 yd. 1 ft. 6 in.

Ex. 17B

1. 12 in.   2. 2 ft. 6 in.   3. 3 ft. 4 in.   4. 3 ft. 8 in.
5. 2 ft. 8½ in.   6. 3 yd. 2 ft.   7. 2 yd. 1 ft. 11 in.
8. 1960 yd.   9. 1 yd. 2 ft. 2 in.   10. 6 yd. 0 ft. 6 in.

Ex. 18A

1. 49 sq. in.   2. 40 sq. in.   3. 99 sq. in.   4. 102 sq. in.   5. 84 sq. in.
6. 33¾ sq. in.   7. 302½ sq. in.   8. 142½ sq. in.   9. 52¼ sq. in.   10. 134½ sq. in.

Ex. 18B

1. 36 sq. in.   2. 63 sq. in.   3. 96 sq. in.   4. 75 sq. in.   5. 69 sq. in.
6. 36¼ sq. in.   7. 315 sq. in.   8. 127½ sq. in.   9. 55¼ sq. in.   10. 168¾ sq. in.

NOTES AND MEMORANDA

Progress Tests

Ex. 19 A

1. How many 1*d*. stamps can I buy with 4*s*. 7*d*.?

2. 2 ft. 5 in. − 1 ft. 9 in.

3. Multiply £7. 8*s*. 4*d*. by 5.

4. Add together 3*s*. 9½*d*. and 8*s*. 7¾*d*.

5. 375 ÷ 25.

6. What is the cost of 481 articles at 8¼*d*. each?

7. Find the cost of 1 dozen at 1*s*. 2¼*d*. each.

8. How much shall I have to pay for 2 score of articles at 3*s*. 6*d*. each?

9. What is the cost of 1 gross of articles at 8*d*. each?

10. Find (*a*) the perimeter, (*b*) the area of a football pitch which is 105 yd. long and 80 yd. wide.

Progress Tests (*continued*)

Ex. 19 B

1. 62 × 300.

2. Take 4*s*. 10½*d*. from ten shillings.

3. 104 × 125.

4. At 7*s*. 3*d*. per dozen, what will be the cost of 1?

5. What will be the cost of 1 if a score cost £7. 15*s*. 0*d*.?

6. 23 × 99.

7. Find the cost of 38 articles at 5*s*. each.

8. How much shall I have to pay for a gross of articles costing 9*d*. each?

9. How much shall I have to pay for 43 three-halfpenny stamps?

10. Find (*a*) the perimeter, (*b*) the area of a rectangular field 108 yd. long and 70 yd. wide.

ANSWERS

Ex. 19 A

1. 55.
2. 8 in.
3. £37. 1s. 8d.
4. 12s. 5¼d.
5. 15.
6. £16. 10s. 8¼d.
7. 14s. 3d.
8. £7. 0s. 0d.
9. £4. 16s. 0d.
10. (a) 370 yd.; (b) 8400 sq. yd.

Ex. 19 B

1. 18,600.
2. 5s. 1½d.
3. 13,000.
4. 7¼d.
5. 7s. 9d.
6. 2277.
7. £9. 10s. 0d.
8. £5. 8s. 0d.
9. 5s. 4½d.
10. (a) 356 yd.; (b) 7560 sq. yd.

NOTES AND MEMORANDA

More Progress Tests

Ex. 20 A

1. 543×7.

2. $14s.\ 6d. + 11s.\ 10\frac{1}{2}d.$

3. How many three-halfpenny stamps can I buy with $6s.\ 7\frac{1}{2}d.$?

4. Find the total cost of 2 dozen at $3d.$ each and 7 at $6d.$ each.

5. 37×101.

6. What will be the cost of 1 if 2 score cost £8?

7. How many pints are there in 2 gal. 3 qt. 1 pt.?

8. What change should I receive from $2s.$ after paying for $1\frac{1}{2}$ lb. cheese at $9d.$ per lb.?

9. How much will 962 articles at $6\frac{1}{2}d.$ cost?

10. The handle of a 10 in. long knife is 2 in. shorter than the blade. What is the length of the blade?

More Progress Tests (*continued*)

Ex. 20 B

1. $524 - 276$.

2. 2 tons 9 cwt. 1 qr. $\times 6$.

3. 48×25.

4. How many pints shall I have left after serving 7 pt. from $2\frac{1}{2}$ gal.?

5. Find the total cost of 1 dozen at $4\frac{3}{4}d.$ each and 17 at $1\frac{1}{2}d.$ each.

6. What will be the cost of 2 score at $7s.\ 6d.$ each?

7. How many inches are there in 1 yd. 2 ft. 7 in.?

8. What must I pay for 24 articles at $7\frac{1}{2}d.$ each?

9. How much change shall I receive from $2s.\ 6d.$ after paying for $1\frac{1}{2}$ lb. of cheese at $10d.$ per lb.?

10. The combined ages of Fred and John is 21. John is 3 years older than Fred. What is Fred's age?

ANSWERS

1. 3801. 2. £1. 6s. 4½d. 3. 53. 4. 9s. 6d. 5. 3737.
6. 4s. 7. 23. 8. 10½d. 9. £26. 1s. 1d. 10. 6 in.

Ex. 20 B

1. 248. 2. 14 tons 15 cwt. 2 qr. 3. 1200. 4. 13 pt. 5. 6s. 10½d.
6. £15. 0s. 0d. 7. 67 in. 8. 15s. 9. 1s. 3d. 10. 9 yr.

NOTES AND MEMORANDA

Prime Factors

Find the prime factors of the following:

	Ex. 21 A				Ex. 21 B	
1.	(a) 6.	(b) 10.		1.	(a) 4	(b) 8.
2.	(a) 14.	(b) 18.		2.	(a) 12.	(b) 15.
3.	(a) 16.	(b) 22.		3.	(a) 20.	(b) 24.
4.	(a) 25.	(b) 30.		4.	(a) 27.	(b) 28.
5.	(a) 21.	(b) 32.		5.	(a) 26.	(b) 33.
6.	(a) 36.	(b) 39.		6.	(a) 40.	(b) 34.
7.	(a) 38.	(b) 45.		7.	(a) 42.	(b) 44.
8.	(a) 50.	(b) 54.		8.	(a) 48.	(b) 49.
9.	(a) 63.	(b) 72.		9.	(a) 52.	(b) 60.
10.	(a) 96.	(b) 104.		10.	(a) 84.	(b) 126.

Least Common Multiple

Ex. 22 A   Ex. 22 B

Find the L.C.M. of: Find the L.C.M. of:

Ex. 22 A	Ex. 22 B
1. 6 and 8.	1. 6 and 9.
2. 9 and 12.	2. 8 and 12.
3. 5 and 6.	3. 4 and 5.
4. 7 and 4.	4. 3 and 8.
5. 2, 3 and 6.	5. 3, 4 and 8.
6. 3, 4 and 5.	6. 4, 6 and 12.
7. 2, 4 and 5.	7. 3, 5 and 6.
8. 2, 6 and 9.	8. 3, 6 and 8.
9. 4, 9 and 12.	9. 6, 8 and 12.
10. 2, 5 and 7.	10. 2, 5 and 9.

ANSWERS

Ex. 21 A

1. (a) 2, 3.   (b) 2, 5.
2. (a) 2, 7.   (b) 2, 3.
3. (a) 2.   (b) 2, 11.
4. (a) 5.   (b) 2, 3, 5.
5. (a) 3, 7.   (b) 2.
6. (a) 2, 3.   (b) 3, 13.
7. (a) 2, 19.   (b) 3, 5.
8. (a) 2, 5.   (b) 2, 3.
9. (a) 3, 7.   (b) 2, 3.
10. (a) 2, 3.   (b) 2, 13.

Ex. 21 B

1. (a) 2.   (b) 2.
2. (a) 2, 3.   (b) 3, 5.
3. (a) 2, 5.   (b) 2, 3.
4. (a) 3.   (b) 2, 7.
5. (a) 2, 13.   (b) 3, 11.
6. (a) 2, 5.   (b) 2, 17.
7. (a) 2, 3, 7.   (b) 2, 11.
8. (a) 2, 3.   (b) 7.
9. (a) 2, 13.   (b) 2, 3, 5.
10. (a) 2, 3, 7.   (b) 2, 3, 7.

Ex. 22 A

1. 24.   2. 36.   3. 30.   4. 28.   5. 6.
6. 60.   7. 20.   8. 18.   9. 36.   10. 70.

Ex. 22 B

1. 18.   2. 24.   3. 20.   4. 24.   5. 24.
6. 12.   7. 30.   8. 24.   9. 24.   10. 90.

NOTES AND MEMORANDA

Reduction to Lowest Terms and Improper Fractions

Ex. 23

A. Reduce to Lowest Terms:

1. (a) $\frac{3}{6}$.   (b) $\frac{12}{16}$.
2. (a) $\frac{5}{15}$.   (b) $\frac{8}{12}$.
3. (a) $\frac{3}{18}$.   (b) $\frac{6}{14}$.
4. (a) $\frac{18}{27}$.   (b) $\frac{35}{42}$.
5. (a) $\frac{36}{45}$.   (b) $\frac{48}{108}$.

Change to mixed numbers:

6. (a) $\frac{5}{3}$.   (b) $\frac{7}{6}$.
7. (a) $\frac{7}{2}$.   (b) $\frac{9}{5}$.
8. (a) $\frac{18}{7}$.   (b) $\frac{19}{8}$.
9. (a) $\frac{43}{10}$.   (b) $\frac{81}{7}$.
10. (a) $\frac{101}{11}$.   (b) $\frac{81}{12}$.

B. Reduce to Lowest Terms:

1. (a) $\frac{4}{8}$.   (b) $\frac{10}{12}$.
2. (a) $\frac{5}{20}$.   (b) $\frac{14}{16}$.
3. (a) $\frac{3}{15}$.   (b) $\frac{8}{22}$.
4. (a) $\frac{22}{33}$.   (b) $\frac{21}{30}$.
5. (a) $\frac{42}{54}$.   (b) $\frac{84}{120}$.

Change to mixed numbers:

6. (a) $\frac{3}{2}$.   (b) $\frac{4}{3}$.
7. (a) $\frac{7}{4}$.   (b) $\frac{8}{5}$.
8. (a) $\frac{15}{4}$.   (b) $\frac{17}{3}$.
9. (a) $\frac{37}{10}$.   (b) $\frac{89}{9}$.
10. (a) $\frac{93}{11}$.   (b) $\frac{104}{12}$.

Mixed Numbers and the Four Rules applied to Fractions

Ex. 23 C

Change to Improper Fractions:

1. (a) $1\frac{3}{4}$.   (b) $1\frac{4}{5}$.
2. (a) $2\frac{3}{7}$.   (b) $4\frac{5}{9}$.
3. (a) $5\frac{3}{8}$.   (b) $6\frac{5}{6}$.
4. (a) $7\frac{1}{9}$.   (b) $4\frac{7}{10}$.
5. (a) $3\frac{8}{11}$.   (b) $7\frac{5}{12}$.

Simplify:

6. $\frac{1}{2}+\frac{1}{3}$.
7. $\frac{1}{8}+\frac{1}{6}$.
8. $\frac{1}{4}-\frac{1}{5}$.
9. $\frac{3}{4}\times\frac{1}{3}$.
10. $\frac{1}{12}\div\frac{1}{3}$.

Ex. 23 D

Change to Improper Fractions:

1. (a) $1\frac{3}{8}$.   (b) $1\frac{5}{6}$.
2. (a) $3\frac{5}{8}$.   (b) $2\frac{7}{9}$.
3. (a) $4\frac{6}{7}$.   (b) $7\frac{4}{5}$.
4. (a) $8\frac{1}{7}$.   (b) $5\frac{9}{10}$.
5. (a) $4\frac{6}{11}$.   (b) $6\frac{7}{12}$.

Simplify:

6. $\frac{1}{3}+\frac{1}{4}$.
7. $\frac{1}{7}+\frac{1}{5}$.
8. $\frac{1}{6}-\frac{1}{7}$.
9. $\frac{4}{5}\times\frac{1}{2}$.
10. $\frac{1}{10}\div\frac{1}{5}$.

ANSWERS

Ex. 23 A

1. (a) $\frac{1}{2}$. (b) $\frac{3}{4}$.
2. (a) $\frac{1}{3}$. (b) $\frac{3}{8}$.
3. (a) $\frac{1}{6}$. (b) $\frac{9}{7}$.
4. (a) $\frac{3}{8}$. (b) $\frac{5}{8}$.
5. (a) $\frac{5}{8}$. (b) $\frac{1}{6}$.
6. (a) $1\frac{3}{8}$. (b) $1\frac{1}{3}$.
7. (a) $3\frac{1}{2}$. (b) $1\frac{4}{5}$.
8. (a) $2\frac{5}{7}$. (b) $2\frac{3}{8}$.
9. (a) $4\frac{5}{10}$. (b) $11\frac{4}{9}$.
10. (a) $9\frac{3}{11}$. (b) $6\frac{3}{4}$.

Ex. 23 B

1. (a) $\frac{1}{2}$. (b) $\frac{5}{8}$.
2. (a) $\frac{1}{4}$. (b) $\frac{7}{8}$.
3. (a) $\frac{1}{5}$. (b) $\frac{4}{11}$.
4. (a) $\frac{3}{8}$. (b) $\frac{7}{10}$.
5. (a) $\frac{7}{6}$. (b) $\frac{7}{10}$.
6. (a) $1\frac{1}{2}$. (b) $1\frac{1}{4}$.
7. (a) $1\frac{3}{4}$. (b) $1\frac{3}{8}$.
8. (a) $3\frac{3}{4}$. (b) $5\frac{4}{5}$.
9. (a) $3\frac{7}{10}$. (b) $9\frac{5}{8}$.
10. (a) $8\frac{5}{11}$. (b) $8\frac{3}{8}$.

Ex. 23 C

1. (a) $\frac{7}{4}$. (b) $\frac{9}{5}$.
2. (a) $\frac{17}{7}$. (b) $\frac{41}{9}$.
3. (a) $\frac{43}{8}$. (b) $\frac{41}{9}$.
4. (a) $\frac{64}{9}$. (b) $\frac{47}{10}$.
5. (a) $\frac{41}{11}$. (b) $\frac{89}{12}$.
6. $\frac{5}{8}$. 7. $\frac{7}{24}$. 8. $\frac{1}{10}$.
9. $\frac{1}{4}$. 10. $\frac{1}{4}$.

Ex. 23 D

1. (a) $\frac{5}{3}$. (b) $\frac{11}{10}$.
2. (a) $\frac{29}{8}$. (b) $\frac{25}{9}$.
3. (a) $\frac{34}{7}$. (b) $\frac{39}{8}$.
4. (a) $\frac{57}{7}$. (b) $\frac{59}{10}$.
5. (a) $\frac{69}{11}$. (b) $\frac{79}{12}$.
6. $\frac{7}{12}$. 7. $\frac{11}{36}$. 8. $\frac{1}{12}$.
9. $\frac{3}{8}$. 10. $\frac{1}{4}$.

NOTES AND MEMORANDA

Fractional Quantities

Money

Ex. 24 A. Find the value of:

1. $\frac{1}{3}$ of 1s.
2. $\frac{3}{8}$ of 1s.
3. $2\frac{1}{6}s$.
4. $\frac{1}{5}$ of 2s. 1d.
5. £$7\frac{5}{8}$.
6. $\frac{11}{40}$ of £1.
7. £$2\frac{1}{3}$.
8. £$5\frac{3}{4}$.
9. $\frac{7}{16}$ of £1.
10. $\frac{5}{9}$ of £2. 5s.

B. Find the value of:

1. $\frac{1}{4}$ of 1s.
2. $\frac{7}{8}$ of 1s.
3. $5\frac{5}{8}s$.
4. $\frac{1}{8}$ of 3s. 6d.
5. £$6\frac{3}{8}$.
6. $\frac{9}{80}$ of £1.
7. £$3\frac{1}{8}$.
8. £$6\frac{1}{5}$.
9. $\frac{5}{12}$ of £1.
10. $\frac{7}{8}$ of £2. 8s.

Measures

Ex. 25

A.
1. $\frac{1}{4}$ of 1 ft. (in in.).
2. $\frac{7}{8}$ of 1 gal. (in pt.).
3. $\frac{1}{4}$ of 1 ml. (in yd.).
4. $4\frac{5}{8}$ lb. (to lb. and oz.).
5. $\frac{2}{15}$ of 1 hr. (min.).
6. $\frac{5}{9}$ of 1 yd. (in in.).
7. $\frac{1}{4}$ of 1 cwt. (in lb.).
8. $\frac{9}{10}$ of 1 ml. (in yd.).
9. $\frac{7}{10}$ of 1 ton (in cwt.).
10. $\frac{3}{8}$ of 1 ml. (in yd.).

B.
1. $\frac{1}{3}$ of 1 ft. (in in.).
2. $\frac{3}{8}$ of 1 gal. (in pt.).
3. $\frac{1}{4}$ of 1 ton (in lb.).
4. $5\frac{7}{8}$ lb. (to lb. and oz.).
5. $\frac{5}{12}$ of 1 hr. (in min.).
6. $\frac{5}{8}$ of 1 yd. (in in.).
7. $\frac{3}{4}$ of 1 qr. (in lb.).
8. $\frac{7}{10}$ of 1 ml. (in yd.).
9. $\frac{13}{20}$ of 1 ton (in cwt.).
10. $\frac{5}{8}$ of 1 ml. (in yd.).

ANSWERS

Ex. 24 A

1. 4*d*.
2. 4½*d*.
3. 2*s*. 2*d*.
4. 5*d*.
5. £7. 12*s*. 6*d*.
6. 5*s*. 6*d*.
7. £2. 6*s*. 8*d*.
8. £5. 15*s*. 0*d*.
9. 8*s*. 9*d*.
10. £1. 5*s*. 0*d*.

Ex. 24 B

1. 3*d*.
2. 10½*d*.
3. 5*s*. 10*d*.
4. 7*d*.
5. £6. 7*s*. 6*d*.
6. 2*s*. 3*d*.
7. £3. 3*s*. 4*d*.
8. £6. 4*s*. 0*d*.
9. 8*s*. 4*d*.
10. £2. 2*s*. 0*d*.

Ex. 25 A

1. 3 in.
2. 7 pt.
3. 440 yd.
4. 4 lb. 10 oz.
5. 8 min.
6. 20 in.
7. 28 lb.
8. 1584 yd.
9. 14 cwt.
10. 660 yd.

Ex. 25 B

1. 4 in.
2. 3 pt.
3. 560.
4. 5 lb. 14 oz.
5. 25 min.
6. 30 in.
7. 21 lb.
8. 1232 yd.
9. 13 cwt.
10. 1100 yd.

NOTES AND MEMORANDA

[18*a*]

Miscellaneous

Ex. 26 A

1. $\frac{1}{2} + \frac{5}{8}$.

2. What fraction of £1 is 4s.?

3. Find the L.C.M. of 2, 5 and 8.

4. $\frac{5}{8} \div \frac{1}{3}$.

5. If a milkman serves 60 pt. from a 10 gal. can, how many gallons will he have left?

6. Find the value of $\frac{5}{8}$ of £1.

7. What weight of sugar will be required to fill 5 packets with 1 lb. 4 oz. each?

8. Change to lb. $\frac{9}{10}$ of 1 ton.

9. Which is the longer and by how much: $\frac{2}{3}$ in. or $\frac{5}{8}$ in.?

10. $\frac{3}{8}$ of £1 + $\frac{2}{3}$ of 1s. (Ans. in s. d.)

Miscellaneous (*continued*)

Ex. 26 B

1. $\frac{1}{3} - \frac{1}{4}$.

2. What fraction of £1 is 5s.?

3. Find the L.C.M. of 3, 4 and 9.

4. $\frac{3}{4} \times \frac{2}{9}$.

5. On an average a cow gives 2 gal. 1 qt. 1 pt. daily. How much does she give in a week?

6. Change to s. d. $\frac{7}{8}$ of £1.

7. The weight of coal in a truck is 9 tons 6 cwt. What weight would be contained in 5 such trucks?

8. How many yards are there in $\frac{3}{4}$ of 1 ml.?

9. Which is the longer and by how much: $\frac{5}{8}$ in. or $\frac{3}{4}$ in.?

10. $\frac{1}{3}$ of £1 + $\frac{5}{8}$ of 1s. (Ans. in s. d.)

ANSWERS

Ex. 26 A

1. $1\frac{1}{4}$.   2. $\frac{1}{8}$.   3. 40.   4. $2\frac{1}{2}$.

5. $2\frac{1}{2}$.   6. 16s. 8d.   7. 6 lb. 4 oz.

8. 2016 lb.   9. $\frac{5}{8}$ in. by $\frac{1}{8}$ in.   10. 8s. 2d.

Ex. 26 B

1. $\frac{1}{12}$.   2. $\frac{1}{4}$.   3. 36.   4. $\frac{1}{8}$.

5. 16 gal. 2 qt. 1 pt.   6. 17s. 6d.   7. 46 tons 10 cwt.

8. 1320 yd.   9. $\frac{3}{4}$ in. by $\frac{1}{8}$ in.   10. 7s. $3\frac{1}{2}$d.

NOTES AND MEMORANDA

Decimals—Conversion

A. Change the following vulgar fractions to decimals:

1. (a) $\frac{3}{10}$.   (b) $\frac{8}{10}$.
2. (a) $\frac{1}{5}$.   (b) $\frac{3}{4}$.
3. (a) $\frac{1}{8}$.   (b) $\frac{7}{8}$.
4. (a) $\frac{47}{100}$.   (b) $\frac{9}{20}$.
5. (a) $\frac{561}{1000}$.   (b) $\frac{10}{25}$.

Change the following decimals to vulgar fractions:

6. (a) ·2.   (b) ·7.
7. (a) ·8.   (b) ·5.
8. (a) ·25.   (b) ·24.
9. (a) ·375.   (b) ·625.
10. (a) ·13.   (b) ·075.

B. Change the following vulgar fractions to decimals:

1. (a) $\frac{7}{10}$.   (b) $\frac{4}{10}$.
2. (a) $\frac{1}{4}$.   (b) $\frac{3}{5}$.
3. (a) $\frac{3}{8}$.   (b) $\frac{5}{8}$.
4. (a) $\frac{51}{100}$.   (b) $\frac{11}{20}$.
5. (a) $\frac{423}{1000}$.   (b) $\frac{8}{25}$.

Change the following decimals to vulgar fractions:

6. (a) ·3.   (b) ·4.
7. (a) ·6.   (b) ·9.
8. (a) ·75.   (b) ·16.
9. (a) ·125.   (b) ·875.
10. (a) ·17.   (b) ·025.

The Four Rules applied to Decimals

A.
1. $2·2 + 3·5$.
2. $·8 + ·7$.
3. $2·6 - 1·9$.
4. $6·38 - 3·75$.
5. $1·35 \times 4$.
6. $3·25 \div 5$.
7. $5·34 \times 20$.
8. $6·25 \div ·125$.
9. $5·7 + ·23 + 1·06$.
10. $8·4 \times ·25$.

B.
1. $3·3 + 4·2$.
2. $·9 + ·5$.
3. $3·2 - 2·6$.
4. $8·46 - 5·82$.
5. $1·26 \times 5$.
6. $3·68 \div 4$.
7. $6·43 \times 20$.
8. $8·75 \div ·125$.
9. $8·24 + 2·5 + ·04$.
10. $5·6 \times ·25$.

ANSWERS

1. (a) ·3.   (b) ·8.
2. (a) ·2.   (b) ·75.
3. (a) ·125.   (b) ·875.
4. (a) ·47.   (b) ·45.
5. (a) ·561.   (b) ·4.
6. (a) $\frac{1}{5}$.   (b) $\frac{7}{10}$.
7. (a) $\frac{4}{5}$.   (b) $\frac{1}{8}$.
8. (a) $\frac{1}{4}$.   (b) $\frac{6}{25}$.
9. (a) $\frac{3}{8}$.   (b) $\frac{5}{8}$.
10. (a) $\frac{13}{100}$.   (b) $\frac{3}{40}$.

Ex. 27 B

1. (a) ·7.   (b) ·4.
2. (a) ·25.   (b) ·6.
3. (a) ·375.   (b) ·625.
4. (a) ·51.   (b) ·55.
5. (a) ·423.   (b) ·32.
6. (a) $\frac{9}{10}$.   (b) $\frac{2}{5}$.
7. (a) $\frac{3}{8}$.   (b) $\frac{9}{10}$.
8. (a) $\frac{3}{4}$.   (b) $\frac{4}{25}$.
9. (a) $\frac{1}{4}$.   (b) $\frac{7}{8}$.
10. (a) $\frac{17}{100}$.   (b) $\frac{1}{40}$.

Ex. 28 A

1. 5·7.   2. 1·5.   3. ·7.   4. 2·63.   5. 5·4.
6. ·65.   7. 106·8.   8. 50.   9. 6·99.   10. 2·1.

Ex. 28 B

1. 7·5.   2. 1·4.   3. ·6.   4. 2·64.   5. 6·3.
6. ·92.   7. 128·6.   8. 70.   9. 10·78.   10. 1·4.

NOTES AND MEMORANDA

Decimal Quantities

Money

Ex. 29 A. Find the value of:

1. ·25 of 1s.
2. ·875 of 1s.
3. ·125 of 1s.
4. ·1 of £1.
5. ·7 of £1.
6. ·2 of £1.
7. ·375 of £1.
8. ·45 of £1.
9. ·075 of £1.
10. ·0625 of £1.

B. Find the value of:

1. ·75 of 1s.
2. ·625 of 1s.
3. ·375 of 1s.
4. ·5 of £1.
5. ·3 of £1.
6. ·8 of £1.
7. ·875 of £1.
8. ·55 of £1.
9. ·025 of £1.
10. ·0125 of £1.

Measures

Ex. 30

A.
1. ·25 of 1 ft. (in in.).
2. ·2 of 1 ton (in cwt.).
3. ·375 of 1 gal. (in pt.).
4. ·8 of 1 hr. (in min.).
5. ·875 of 1 lb. (in oz.).
6. ·4 of 1 ml. (in yd.).
7. ·025 of 1 ton (in lb.).
8. ·35 of 1 hr. (in min.).
9. ·6 of 5 yd. (in ft.).
10. ·75 of 2 lb. (in oz.).

B.
1. ·75 of 1 yd. (in in.).
2. ·5 of 1 hr. (in min.).
3. ·625 of 1 gal. (in pt.).
4. ·7 of 1 ton (in cwt.).
5. ·125 of 1 lb. (in oz.).
6. ·6 of 1 ton (in lb.).
7. ·025 of 1 ml. (in yd.).
8. ·45 of 1 hr. (in min.).
9. ·4 of 5 gal. (in qt.).
10. ·25 of 6 lb. (in oz.).

ANSWERS

Ex. 29 A

1. 3*d.*	2. 10½*d.*	3. 1½*d.*	4. 2*s.*	5. 14*s.*
6. 4*s.*	7. 7*s.* 6*d.*	8. 9*s.*	9. 1*s.* 6*d.*	10. 1*s.* 3*d.*

Ex. 29 B

1. 9*d.*	2. 7½*d.*	3. 4½*d.*	4. 10*s.*	5. 6*s.*
6. 16*s.*	7. 17*s.* 6*d.*	8. 11*s.*	9. 6*d.*	10. 3*d.*

Ex. 30 A

1. 3 in.	2. 4 cwt.	3. 3 pt.	4. 48 min.	5. 14 oz.
6. 704 yd.	7. 56 lb.	8. 21 min.	9. 9 ft.	10. 24 oz.

Ex. 30 B

1. 27 in.	2. 30 min.	3. 5 pt.	4. 14 cwt.	5. 2 oz.
6. 1344 lb.	7. 44.	8. 27 min.	9. 8 qt.	10. 24 oz.

NOTES AND MEMORANDA

Miscellaneous Exercises

Ex. 31 A

1. Write $\frac{17}{100}$ as a decimal.

2. Reduce to its lowest terms $\frac{16}{24}$.

3. Change $\frac{121}{9}$ to a mixed number.

4. Change $2\frac{5}{8}$ to an improper fraction.

5. $\frac{1}{2} + \frac{1}{5}$.

6. Each week a young man saves $\frac{1}{7}$th of his wages which are £1. 9s. 2d. How much does he save in a fortnight?

7. $£\frac{1}{8} + \cdot25$ of 1s. Give your answer in s. d.

8. What fraction of £1 is 7s. 6d.?

9. $6\cdot17 \times \cdot3$.

10. If I start out with £·5 and spend £$\frac{1}{8}$ at one shop and ·875s. at another how much shall I have left?

Miscellaneous Exercises *(continued)*

Ex. 31 B

1. Write ·12 as a vulgar fraction, in its lowest terms.

2. $2\cdot9 + 5\cdot7$.

3. Change $\frac{143}{8}$ to a mixed number.

4. $\frac{1}{8} - \frac{1}{9}$.

5. Change $3\frac{4}{7}$ to an improper fraction.

6. 1 ton 8 cwt. 2 qr. of potatoes is to be equally divided among 6 families. What weight of potatoes will each family receive?

7. $£\cdot25 + \frac{3}{4}$ of 1s. Give your answer in s. d.

8. What decimal part of 1 ton is 8 cwt.?

9. $1\cdot65 \div \cdot5$.

10. One cow gave 2·5 gal. of milk and another 1·875 gal. What was the total amount (in gal., qt., pt.) given by the 2 cows?

ANSWERS

Ex. 31 A

1. ·17. 2. $\frac{2}{3}$. 3. $13\frac{1}{3}$. 4. $2\frac{1}{8}$. 5. $\frac{7}{10}$.

6. 8s. 4d. 7. 4s. 3d. 8. $\frac{3}{8}$. 9. 1·851. 10. 2s. $5\frac{1}{2}d$.

Ex. 31 B

1. $\frac{3}{25}$. 2. 8·6. 3. $17\frac{7}{8}$. 4. $\frac{1}{72}$. 5. $\frac{25}{7}$.

6. 4 cwt. 3 qr. 7. 5s. $4\frac{1}{2}d$. 8. ·4. 9. 3·3. 10. 4 gal. 1 qt. 1 pt

NOTES AND MEMORANDA

Progress Tests

1. 1 lb. 13 oz. + 2 lb. 8 oz.

2. $3s. 4\frac{1}{2}d. \times 5$.

3. How many pence are there in $9s. 10d.$?

4. Find the cost of 2 dozen articles at $4\frac{1}{4}d.$ each.

5. How much shall I have to pay for 1 gross of articles at $6\frac{1}{2}d.$ each?

6. What change shall I receive from a $10s.$ note after paying for $3\frac{1}{2}$ lb. of beef at $1s. 2d.$ per lb.?

7. Write down the prime factors of 56.

8. Change $4\frac{7}{11}$ to an improper fraction.

9. $27\cdot625 - 19\cdot789$.

10. $£\frac{3}{8} - 10\cdot375s.$ Give your answer in $s. d.$

Progress Tests *(continued)*

1. 2 yd. 1 ft. 8 in. − 1 yd. 2 ft. 10 in.

2. 2 tons 9 cwt. 2 qr. ÷ 9.

3. 76×25.

4. Find the cost of a score of articles at $5s. 9d.$ each.

5. How much shall I have to pay for 479 articles at $7\frac{3}{4}d.$ each?

6. What will be the total cost of $1\frac{1}{2}$ lb. cheese at $1s.$ per lb., and 2 lb. butter at $1s. 5d.$ per lb.?

7. Reduce to its lowest terms $\frac{18}{54}$.

8. Change $\frac{113}{12}$ to a mixed number.

9. $3\cdot07 + 1\cdot25 + \cdot43$.

10. $£\frac{4}{5} - 12\cdot625s.$ Give your answer in $s. d.$

ANSWERS

Ex. 32 A

1. 4 lb. 5 oz. 2: 16s. 10½d. 3. 118. 4. 8s. 6d. 5. £3. 18s. 0d.

6. 5s. 11d. 7. 2, 7. 8. $\frac{9}{11}$. 9. 7·836. 10. 1s. 7½d.

Ex. 32 B

1. 1 ft. 10 in. 2. 5 cwt. 2 qr. 3. 1900. 4. £5. 15s. 0d. 5. £15. 9s. 4¼d.

6. 4s. 4d. 7. $\frac{1}{3}$. 8. $9\frac{5}{12}$. 9. 4·75. 10. 3s. 4½d.

NOTES AND MEMORANDA

More Progress Tests

Ex. 33 A

1. £4. 16s. 7½d. − £1. 12s. 8d.
2. 65 × 300.
3. If 2 score of articles cost £15. 10s. 0d. What is the cost of 1?
4. What will be the total cost of 2 lb. of bacon at 1s. 1d. per lb., and 1½ lb. of butter at 1s. 4d. per lb.?
5. How much shall I have to pay for 959 articles at 8¼d. each?
6. ¾ + ⅛.
7. If from a can containing 4½ gal. of milk the milkman sells 25 pt., how many pints will he have left?
8. Find (a) the perimeter, and (b) the area of a rectangle 10 in. long and 4½ in. wide.
9. 3·28 ÷ ·8.
10. £·625 + ·75 of 1s. Give your answer in s. d.

More Progress Tests (*continued*)

Ex. 33 B

1. £3. 18s. 9d. + £2. 15s. 6½d.
2. Change 94 pence to s. d.
3. If 3 dozen articles cost 6s. What will be the cost of 1?
4. How much change shall I receive from a 10s. note after paying for 2½ lb. lamb at 1s. 7d. per lb.?
5. If 1 gross of articles cost £4. 16s., what is the cost of 1 of them?
6. ⅔ + ⅐.
7. What is the total number of days in the months of June, July, and August?
8. Find (a) the perimeter, and (b) the area of a rectangle 8 in. long and 5½ in. wide.
9. 5·27 × 20.
10. £·375 + ⅝ of 1s. Give your answer in s. d.

[24]

ANSWERS

Ex. 33 A

1. £3. 3s. 11½d.
2. 19,500.
3. 7s. 9d.
4. 4s. 2d.
5. £32. 19s. 3¼d.
6. 1¹⅓.
7. 11 pt.
8. (a) 2 ft. 5 in.   (b) 45 sq. in.
9. 4·1.
10. 13s. 3d.

Ex. 33 B

1. £6. 14s. 3½d.
2. 7s. 10d.
3. 2d.
4. 6s. 0½d.
5. 8d.
6. 1⁷⁄₁₁.
7. 92.
8. (a) 2 ft. 3 in.   (b) 44 sq. in.
9. 105·4.
10. 8s. 4d.

NOTES AND MEMORANDA

Averages

Ex. 34

A. Find the averages of the following:

1. 3; 4; 5.
2. 2; 4; 6; 8.
3. 1s. 4½d.; 4s. 7½d.
4. 2 ft. 7 in.; 3 ft. 3 in.
5. 3 hr. 10 min.; 4 hr. 20 min.
6. 7; 9; 12; 8.
7. 2·6; 1·8.
8. 4·06; ·54.
9. What is the average speed in m.p.h. of a woman who walks 13½ ml. in 3 hr.?
10. At an average speed of 11 m.p.h. how long will it take a cyclist to cover 209 ml.?

B. Find the averages of the following:

1. 2; 4; 3.
2. 3; 5; 7; 9.
3. 2s. 6½d.; 4s. 5¼d.
4. 1 ft. 8 in.; 3 ft. 3 in.
5. 2 hr. 5 min.; 5 hr. 25 min.
6. 4; 8; 14; 6.
7. 3·5; 1·7.
8. 2·07; ·73.
9. What is the average speed in m.p.h. of a man who walks 18 ml. in 4 hr.?
10. At an average speed of 12 m.p.h. how long will it take a cyclist to cover 216 ml.?

Times

Ex. 35 A

Ex. 35 B

Find the number of hr. and min. from the first to the second time:

	A	B		A	B
1.	8.45 a.m.	10.52 a.m.	1.	7.37 a.m.	11.46 a.m.
2.	7.23 a.m.	11.49 a.m.	2.	8.28 a.m.	10.52 a.m.
3.	9.52 a.m.	11.27 a.m.	3.	10.49 a.m.	11.56 a.m.
4.	11.25 a.m.	2.30 p.m.	4.	10.24 a.m.	2.35 p.m.
5.	10.14 a.m.	3.51 p.m.	5.	9.17 a.m.	2.56 p.m.
6.	7.34 a.m.	9.43 p.m.	6.	8.32 a.m.	10.46 p.m.
7.	11.38 a.m.	4.22 p.m.	7.	11.55 a.m.	4.15 p.m.
8.	9.56 a.m.	2.17 p.m.	8.	10.45 a.m.	3.28 p.m.
9.	2.29 p.m.	10.16 p.m.	9.	3.27 p.m.	11.12 p.m.
10.	5.35 p.m.	1.27 a.m.	10.	6.25 p.m.	2.14 a.m.

ANSWERS

1. 4.	2. 5.	3. 3*s*.	4. 2 ft. 11 in.	5. 3 hr. 45 min.
6. 9.	7. 2·2.	8. 2·3.	9. $4\frac{1}{2}$ m.p.h.	10. 19 hr.

Ex. 34 B

1. 3.	2. 6.	3. 3*s*. 6*d*.	4. 2 ft. $5\frac{1}{2}$ in.	5. 3 hr. 45 min.
6. 8.	7. 2·6.	8. 1·4.	9. $4\frac{1}{2}$ m.p.h.	10. 18 hr.

Ex. 35 A	**Ex. 35 B**
1. 2 hr. 7 min.	1. 4 hr. 9 min.
2. 4 hr. 26 min.	2. 2 hr. 24 min.
3. 1 hr. 35 min.	3. 1 hr. 7 min.
4. 3 hr. 5 min.	4. 4 hr. 11 min.
5. 5 hr. 37 min.	5. 5 hr. 39 min.
6. 14 hr. 9 min.	6. 14 hr. 14 min.
7. 4 hr. 44 min.	7. 4 hr. 20 min.
8. 4 hr. 21 min.	8. 4 hr. 43 min.
9. 7 hr. 47 min.	9. 7 hr. 45 min.
10. 7 hr. 52 min.	10. 7 hr. 49 min.

NOTES AND MEMORANDA

Percentages

A. Change the following vulgar fractions to percentages:

1. (a) $\frac{1}{10}$.   (b) $\frac{9}{10}$.
2. (a) $\frac{1}{4}$.   (b) $\frac{3}{5}$.
3. (a) $\frac{1}{8}$.   (b) $\frac{9}{20}$.
4. (a) $\frac{1}{2}$.   (b) $\frac{1}{40}$.
5. (a) $\frac{4}{5}$.   (b) $\frac{6}{25}$.

Change the following decimals to percentages:

6. (a) ·2.   (b) ·8.
7. (a) ·45.   (b) ·75.
8. (a) 1·68.   (b) 1·97.
9. (a) ·07.   (b) ·025.
10. (a) ·125.   (b) ·875.

B. Change the following vulgar fractions to percentages:

1. (a) $\frac{3}{10}$.   (b) $\frac{7}{10}$.
2. (a) $\frac{1}{5}$.   (b) $\frac{3}{4}$.
3. (a) $\frac{1}{20}$.   (b) $\frac{7}{8}$.
4. (a) $\frac{1}{3}$.   (b) $\frac{1}{50}$.
5. (a) $\frac{2}{5}$.   (b) $\frac{8}{25}$.

Change the following decimals to percentages:

6. (a) ·4.   (b) ·6.
7. (a) ·25.   (b) ·85.
8. (a) 1·36.   (b) 1·75.
9. (a) ·05.   (b) ·075.
10. (a) ·375.   (b) ·625.

Percentages (*continued*)

A. Change the following percentages to vulgar fractions:

1. (a) 20%.   (b) 80%.
2. (a) 45%.   (b) 75%.
3. (a) 5%.   (b) $33\frac{1}{3}$%.
4. (a) $7\frac{1}{2}$%.   (b) $42\frac{1}{2}$%.
5. (a) $12\frac{1}{2}$%.   (b) 85%.

Change the following percentages to decimals:

6. (a) 10%.   (b) 90%.
7. (a) 55%.   (b) 67%.
8. (a) $87\frac{1}{2}$%.   (b) 29%.
9. (a) 2%.   (b) $7\frac{1}{2}$%.
10. (a) $32\frac{1}{2}$%.   (b) $77\frac{1}{2}$%.

B. Change the following percentages to vulgar fractions:

1. (a) 40%.   (b) 60%.
2. (a) 35%.   (b) 25%.
3. (a) 15%.   (b) 50%.
4. (a) $2\frac{1}{2}$%.   (b) $47\frac{1}{2}$%.
5. (a) $37\frac{1}{2}$%.   (b) 65%.

Change the following percentages to decimals:

6. (a) 30%.   (b) 70%.
7. (a) 95%.   (b) 32%.
8. (a) $62\frac{1}{2}$%.   (b) 87%.
9. (a) 6%.   (b) $2\frac{1}{2}$%.
10. (a) $22\frac{1}{2}$%.   (b) $97\frac{1}{2}$%.

ANSWERS

Ex. 36A

1. (a) 10 %. (b) 90 %.
2. (a) 25 %. (b) 60 %.
3. (a) $12\frac{1}{2}$ %. (b) 45 %.
4. (a) 50 %. (b) $2\frac{1}{4}$ %.
5. (a) 80 %. (b) 24 %.
6. (a) 20 %. (b) 80 %.
7. (a) 45 %. (b) 75 %.
8. (a) 168 %. (b) 197 %.
9. (a) 7 %. (b) $2\frac{1}{2}$ %.
10. (a) $12\frac{1}{2}$ %. (b) $87\frac{1}{2}$ %.

Ex. 36B

1. (a) 30 %. (b) 70 %.
2. (a) 20 %. (b) 75 %.
3. (a) 5 %. (b) $87\frac{1}{2}$ %.
4. (a) $33\frac{1}{3}$ %. (b) 2 %.
5. (a) 40 %. (b) 32 %.
6. (a) 40 %. (b) 60 %.
7. (a) 25 %. (b) 85 %.
8. (a) 136 %. (b) 175 %.
9. (a) 5 %. (b) $7\frac{1}{2}$ %.
10. (a) $37\frac{1}{2}$ %. (b) $62\frac{1}{2}$ %.

Ex. 37A

1. (a) $\frac{1}{8}$. (b) $\frac{4}{5}$.
2. (a) $\frac{9}{20}$. (b) $\frac{3}{4}$.
3. (a) $\frac{1}{20}$. (b) $\frac{1}{3}$.
4. (a) $\frac{3}{40}$. (b) $1\frac{7}{10}$.
5. (a) $\frac{1}{4}$. (b) $1\frac{7}{10}$.
6. (a) ·1. (b) ·9.
7. (a) ·55. (b) ·67.
8. (a) ·875. (b) ·29.
9. (a) ·02. (b) ·075.
10. (a) ·325. (b) ·775.

Ex. 37B

1. (a) $\frac{2}{5}$. (b) $\frac{3}{8}$.
2. (a) $\frac{7}{20}$. (b) $\frac{1}{4}$.
3. (a) $\frac{3}{20}$. (b) $\frac{1}{8}$.
4. (a) $\frac{1}{40}$. (b) $1\frac{9}{10}$.
5. (a) $\frac{3}{8}$. (b) $1\frac{3}{20}$.
6. (a) ·3. (b) ·7.
7. (a) ·95. (b) ·32.
8. (a) ·625. (b) ·87.
9. (a) ·06. (b) ·025.
10. (a) ·225. (b) ·975.

NOTES AND MEMORANDA

Percentage Quantities. Money

A. Find the value of:

1. 25 % of 1s.
2. 87½ % of 1s.
3. 66⅔ % of 1s.
4. 40 % of £1.
5. 75 % of £1.
6. 37½ % of £1.
7. 15 % of £1.
8. 2½ % of £1.
9. 32½ % of £1.
10. 77½ % of £1.

B. Find the value of:

1. 50 % of 1s.
2. 62½ % of 1s.
3. 33⅓ % of 1s.
4. 60 % of £1.
5. 25 % of £1.
6. 12½ % of £1.
7. 35 % of £1.
8. 5 % of £1.
9. 42½ % of £1.
10. 57½ % of £1.

Miscellaneous Exercises on Percentages

Ex. 39

A. Find the amounts of the following percentages:

1. 12 % of 100.
2. 50 % of 22.
3. 80 % of 200.
4. 5 % of 40.
5. 12½ % of 2 gal. (in pt.).
6. 87½ % of 1 lb. (in oz.).
7. 10 % of 1 ton (in lb.).
8. Find the sum of 20 % of £1 and 62½ % of 1s.
9. From 25 % of £1 take 75 % of 1s.
10. Add together 22½ % of £1 and 15 % of £1.

B. Find the amounts of the following percentages:

1. 18 % of 100.
2. 25 % of 32.
3. 20 % of 300.
4. 2½ % of 80.
5. 37½ % of 1 gal. (in pt.).
6. 62½ % of 2 lb. (in oz.).
7. 10 % of 1 ml. (in yd.).
8. Add together 60 % of £1 and 37½ % of 1s.
9. Subtract 25 % of 1s. from 75 % of £1.
10. Find the sum of 42½ % of £1 and 25 % of £1.

ANSWERS

Ex. 38 A

1. 3*d*.
2. 10½*d*.
3. 8*d*.
4. 8*s*.
5. 15*s*.
6. 7*s*. 6*d*.
7. 3*s*.
8. 6*d*.
9. 6*s*. 6*d*.
10. 15*s*. 6*d*.

Ex. 38 B

1. 6*d*.
2. 7½*d*.
3. 4*d*.
4. 12*s*.
5. 5*s*.
6. 2*s*. 6*d*.
7. 7*s*.
8. 1*s*.
9. 8*s*. 6*d*.
10. 11*s*. 6*d*.

Ex. 39 A

1. 12.
2. 11.
3. 160.
4. 2.
5. 2 pt.
6. 14 oz.
7. 224 lb.
8. 4*s*. 7½*d*.
9. 4*s*. 3*d*.
10. 7*s*. 6*d*.

Ex. 39 B

1. 18.
2. 8.
3. 60.
4. 2.
5. 3 pt.
6. 20 oz.
7. 176 yd.
8. 12*s*. 4½*d*.
9. 14*s*. 9*d*.
10. 13*s*. 6*d*.

NOTES AND MEMORANDA

Proportion

1. At 3 for a shilling, what is the cost of 6?
2. If 9 articles cost 2s. 3d., what would be the cost of 1?
3. How much should I have to pay for 5 articles at 7d. each?
4. If 20 pencils cost 2s. 6d., what would be the cost of 50?
5. 3 books cost 4s. 6d. What will 5 cost?
6. If you can buy potatoes at 7 lb. for 1s., what will you have to pay for 1 cwt. of them?
7. If 7 yd. of cloth cost £1. 15s., how much shall I have to pay for 4 yd.?
8. The cost of 960 articles is £4. What will be the cost of half-a-dozen such articles?
9. If cooking eggs are 9 for 1s., what will be the cost of 1½ dozen?
10. Travelling at an average speed of 35 miles per hour how long will it take a motorist to cover 7 miles?

Proportion *(continued)*

1. At 8 for 6d., what is the cost of 4?
2. If 7 articles cost 1s. 9d., what would be the cost of 1?
3. How much shall I have to pay for 4 articles at 9d. each?
4. If 30 penholders cost 2s. 6d., what would be the cost of 70?
5. 4 lb. of butter cost 5s. What will be the cost of 6 lb.?
6. ½ cwt. of potatoes were bought for 4s. At this rate what would be the cost of 7 lb.?
7. If 8 yd. of material cost £1. 12s., what would be the cost of 5 yd. of this material?
8. The cost of 480 articles is £3. What will be the cost of a dozen such articles?
9. If buns are 8 for 6d., what will be the cost of 2 dozen?
10. How long will it take a motor-cyclist to cover 9 miles if he travels at an average speed of 54 miles per hour?

ANSWERS

Ex. 40 A

1. 2*s*.	2. 3*d*.	3. 2*s*. 11*d*.	4. 6*s*. 3*d*.	5. 7*s*. 6*d*.
6. 16*s*.	7. £1.	8. 6*d*.	9. 2*s*.	10. 12 min.

Ex. 40 B

1. 3*d*.	2. 3*d*.	3. 3*s*.	4. 5*s*. 10*d*.	5. 7*s*. 6*d*.
6. 6*d*.	7. £1.	8. 1*s*. 6*d*.	9. 1*s*. 6*d*.	10. 10 min.

NOTES AND MEMORANDA

Mensuration

A. Find the perimeters of rectangles having the following dimensions:

1. 3 in. by 9 in.
2. 6 in. by 12 in.
3. 5 in. by 1 ft. 6 in.
4. $4\frac{1}{2}$ in. by 10 in.
5. $6\frac{1}{2}$ in. by 1 ft. $2\frac{1}{2}$ in.

Find the areas of rectangles having the following dimensions:

Breadth	Length
6. 2 in.	7 in.
7. 7 in.	9 in.
8. 6 in.	1 ft. 3 in.
9. $8\frac{1}{2}$ in.	10 in.
10. $9\frac{1}{2}$ in.	1 ft. 2 in.

B. Find the perimeters of rectangles having the following dimensions:

1. 4 in. by 8 in.
2. 7 in. by 11 in.
3. 7 in. by 1 ft. 1 in.
4. $3\frac{1}{2}$ in. by 12 in.
5. $5\frac{1}{2}$ in. by 1 ft. $4\frac{1}{2}$ in.

Find the areas of rectangles having the following dimensions:

Breadth	Length
6. 3 in.	6 in.
7. 8 in.	11 in.
8. 5 in.	1 ft. 2 in.
9. 6 in.	$10\frac{1}{2}$ in.
10. $4\frac{1}{2}$ in.	1 ft. 6 in.

Algebra

A.
1. $8d + 10d + 11d$.
2. $17y + 28y$.
3. $2a \times 4b$.
4. $5x \times 6x$.
5. $2a \times 4a \times 6a$.
6. $16m \div 4$.
7. $3a^2 \div 9a$.
8. If $4x = 24$, what is the value of x?
9. If $3x - 2 = 13$, what is the value of x?
10. Find the area of a field $4p$ yd. long and $3t$ yd. wide.

B.
1. $7d + 9d + 11d$.
2. $19x + 34x$.
3. $3b \times 3a$.
4. $3y \times 8y$.
5. $3b \times 5b \times 4b$.
6. $15m \div 3$.
7. $2b^2 \div 8b$.
8. If $6y = 18$, what is the value of y?
9. If $4y - 3 = 17$, what is the value of y?
10. Find the area of a field $5r$ yd. long and $4s$ yd. wide.

ANSWERS

Ex. 41 A

1. 2 ft.   2. 3 ft.   3. 3 ft. 10 in.   4. 2 ft. 5 in.   5. 3 ft. 6 in.

6. 14 sq. in.   7. 63 sq. in.   8. 90 sq. in.   9. 85 sq. in.   10. 133 sq. in.

Ex. 41 B

1. 2 ft.   2. 3 ft.   3. 3 ft. 4 in.   4. 2 ft. 7 in.   5. 3 ft. 8 in.

6. 18 sq. in.   7. 88 sq. in.   8. 70 sq. in.   9. 63 sq. in.   10. 81 sq. in.

Ex. 42 A

1. $29d$.   2. $45y$.   3. $8ab$.   4. $30x^3$.   5. $48a^3$.

6. $4m$.   7. $\dfrac{a}{3}$.   8. $x=6$.   9. $x=5$.   10. $12pt$ sq. yd.

Ex. 42 B

1. $27d$.   2. $53x$.   3. $9ab$.   4. $24y^2$.   5. $60b^3$.

6. $5m$.   7. $\dfrac{b}{4}$.   8. $y=3$.   9. $y=5$.   10. $20rs$ sq. yd.

NOTES AND MEMORANDA

Sales' Reductions

Ex. 43 A. In the following questions you have the listed price of articles and the amount of their reduction during sales. Find the actual amount you would have to pay for the articles during the sales.

1. Girls' gloves at 5s. per pair. Reduction of 1d. in 1s.
2. Boys' shoes at 8s. 6d. per pair. Reduction of 1d. in 1s.
3. Men's suits at £4 each. Reduction of 2s. in £1.
4. Women's raincoats at £2. 10s. each. Reduction of 2s. 6d. in £1.
5. Men's hats at 10s. each. Reduction of 20%.
6. Girls' shoes at 12s. 6d. per pair. Reduction of 20%.
7. Women's umbrellas at 16s. each. Reduction of 25%.
8. Men's overcoats at £4 each. Reduction of 15%.
9. Boys' raincoats at £1 each. Reduction of 12½%.
10. Men's gauntlets at 15s. per pair. Reduction of 10%.

Ex. 43 B. In the following questions you have the listed price of articles and the amount of their reduction during sales. Find the actual amount you would have to pay for the articles during the sales.

1. Boys' gloves at 6s. per pair. Reduction of 1d. in 1s.
2. Girls' shoes at 7s. 6d. per pair. Reduction of 1d. in 1s.
3. Men's overcoats at £3 each. Reduction of 2s. in £1.
4. Women's raincoats at £1. 10s. each. Reduction of 2s. 6d. in £1.
5. Men's hats at 15s. each. Reduction of 20%.
6. Boys' shoes at 7s. 6d. per pair. Reduction of 20%.
7. Men's umbrellas at £1 each. Reduction of 25%.
8. Men's suits at £5 each. Reduction of 10%.
9. Boys' raincoats at 16s. each. Reduction of 12½%.
10. Women's gauntlets at 10s. per pair. Reduction of 15%.

ANSWERS

NOTES AND MEMORANDA

Progress Tests

1. $3s.\ 2\frac{1}{2}d. + 4s.\ 9\frac{1}{2}d.$

2. Change 70d. to shillings and pence.

3. 37×101.

4. If one article costs $3\frac{1}{2}d.$, what will be the cost of 961 such articles?

5. Find the cost of 18 articles at 4d. each.

6. Find (a) the perimeter, (b) the area of a rectangle which is 10 in. wide and 1 ft. 2 in. long.

7. Reduce $\frac{24}{32}$ to its lowest terms.

8. Add together £$\frac{1}{4}$, £·75, and 50 % of £1.

9. Of 80 pupils 10 were absent. What percentage of them were present?

10. If tennis balls cost 5s. for half-a-dozen how much shall I have to pay for 4 of them?

Progress Tests (*continued*)

1. Take 2 lb. 12 oz. from 4 lb. 7 oz.

2. $275 \div 25$.

3. What is the cost of 2 dozen articles at $2\frac{1}{4}d.$ each?

4. If the cost of one article is 5d., what will be the cost of 1 gross of such articles?

5. What change shall I receive from 10s. after paying for $3\frac{1}{2}$ lb. of cheese at 1s. per lb.?

6. If $5x = 65$, what is the value of x?

7. Change $6\frac{3}{7}$ to an improper fraction.

8. Multiply $\frac{2}{5}$ by ·25 and give your answer as a percentage.

9. Of 200 eggs 20 % were broken. How many of them remained unbroken?

10. If penknives are 4s. per half-a-dozen, how much shall I have to pay for 4 of them?

ANSWERS

Ex. 44 A	Ex. 44 B
1. 8*s*.	1. 1 lb. 11 oz.
2. 5*s*. 10*d*.	2. 11.
3. 3737.	3. 4*s*. 6*d*.
4. £14. 0*s*. 3½*d*.	4. £3.
5. 6*s*.	5. 6*s*. 6*d*.
6. (*a*) 4 ft. (*b*) 140 sq. in.	6. *x* = 13.
7. ¾.	7. ⁴⁴⁄₇.
8. £1. 10*s*. 0*d*.	8. 10 %.
9. 87½ %.	9. 160.
10. 3*s*. 4*d*.	10. 2*s*. 8*d*.

NOTES AND MEMORANDA

Final Tests

1. 56×25.

2. What is the cost of 2 score of articles at 6s. 6d. each?

3. At 1s. 4d. per lb., what will be the cost of a joint of beef weighing 1 lb. 9 oz.?

4. What is the total cost of 2 lb. cheese at 7d. per lb. and 8 eggs at $1\frac{1}{2}d$. each?

5. $\frac{2}{3}$ of 6s. $+\frac{1}{8}$ of 2s.

6. $4 \cdot 73 - 2 \cdot 18$.

7. If $7x = 63$, what is the value of x?

8. Find the average of 5, 3, 10.

9. A shopkeeper bought a number of articles at 4s. each. At what price will he have to sell them in order to make a profit of $12\frac{1}{2}\%$?

10. A train departs at 10.25 a.m. and arrives at its destination at 2.55 p.m. How long was taken over the journey?

Final Tests (*continued*)

1. $3900 \div 30$.

2. How many $1\frac{1}{2}d$. stamps can I buy with 3s.?

3. If I have to pay 8s. for 2 dozen articles, how much do I pay for each one of them?

4. How much change shall I have from 5s. after paying for 2 lb. of cheese at 8d. per lb.?

5. $\cdot 25$ of 4s. $+ 50\%$ of 2s. 6d.

6. $\frac{3}{4} \times \frac{8}{9}$.

7. Find (*a*) the perimeter, and (*b*) the area of a rectangle 3 in. wide by $6\frac{1}{2}$ in. long.

8. Change $37\frac{1}{2}\%$ of £2 to s. d.

9. If 7 articles cost 2s. 4d., what will 2 of them cost?

10. How many minutes and hours are there between 9.36 a.m. and 2.47 p.m.?

ANSWERS

Ex. 45 A

1. 1400.
2. £13.
3. 2s. 1d.
4. 2s. 2d.
5. 4s. 3d.
6. 2·55
7. $x = 9$.
8. 6.
9. 4s. 6d.
10. 4 hr. 30 min.

Ex. 45 B

1. 130.
2. 24.
3. 4d.
4. 3s. 8d.
5. 2s. 3d.
6. $\frac{2}{3}$.
7. (a) 1 ft. 7 in.   (b) $19\frac{1}{2}$ sq. in.
8. 15s.
9. 8d.
10. 5 hr. 11 min.

NOTES AND MEMORANDA

PART TWO

Miscellaneous Exercises

Ex. 1A

1. Add together 9*d*., 4*d*., 7*d*., 6*d*., and give your answer in shillings and pence.

2. How many are there in 9½ dozen?

3. What would be the cost of 6 books at 1*s*. 5*d*. each?

4. How many hours and minutes are there from 9 a.m. till 12.15 p.m.?

5. Take 2*s*. 7½*d*. from 2 florins.

6. 11 × 200.

7. How much change shall I receive from a 10*s*. note after buying 1½ lb. of butter at 1*s*. 4*d*. per lb.?

8. Take three-quarters of 1 dozen from three-quarters of 1 score.

9. What will be the cost of 126 articles at 3*s*. 4*d*. each?

10. If *x* sixpences = £1. 5*s*. 0*d*., what number does *x* represent?

Miscellaneous Exercises (*continued*)

Ex. 1 B

1. Take 37 pence from 81 pence and give your answer in shillings and pence.

2. How many are there in 8½ dozen?

3. What would be the cost of 4 lb. of beef at 1*s*. 10½*d*. per lb.?

4. How many hours and minutes are there from 10.15 a.m. to 12.30 p.m.?

5. Add together 2*s*. 9½*d*. and 5*s*. 5½*d*.

6. 13 × 200.

7. What change shall I receive from half-a-crown after paying for 2½ lb. of cheese at 8*d*. per lb.?

8. Take half a score from half a gross.

9. How much shall I have to pay for 1 gross of articles at 2*s*. 6*d*. each?

10. If *y* threepenny pieces = 8*s*. 6*d*., what number does *y* represent?

ANSWERS

Ex. 1 A

1. 2*s*. 2*d*.   2. 114.   3. 8*s*. 6*d*.   4. 3 hr. 15 min.   5. 1*s*. 4½*d*.

6. 2200.   7. 8*s*.   8. 6.   9. £21.   10. $x = 50$.

Ex. 1 B

1. 3*s*. 8*d*.   2. 102.   3. 7*s*. 6*d*.   4. 2 hr. 15 min.   5. 8*s*. 3*d*.

6. 2600.   7. 10*d*.   8. 62.   9. £18.   10. $y = 34$.

NOTES AND MEMORANDA

Addition and Subtraction

Ex. 2 A

1. $57 + 18 + 24$.
2. Take £1. 15s. 9¾d. from £2.
3. Add together 3 tons 12 cwt. 3 qr. and 2 tons 19 cwt. 1 qr.
4. $981 - 67$.
5. From 6 gal. 3 qt. 1 pt. take 24 pt.
6. From a piece of cloth 2½ yd. long a 2 ft. length is cut. What length is left?
7. 4s. 7½d. + 9s. 10¾d.
8. $5 + a + 11 = 21$. What number does a represent?
9. I start out with £1 and buy 2 articles costing 2s. 7d. and 4s. 6d. respectively. How much money have I left?
10. ½ gross + 1½ score.

Ex. 2 B

1. $65 + 19 + 32$.
2. From £3 take £1. 18s. 6½d.
3. Add together 17 gal. 3 qt. 1 pt. and 6 gal. 0 qt. 1 pt.
4. $873 - 59$.
5. From 3 yd. 2 ft. 5 in. take 50 in.
6. A milkman starts out with 5½ gal. of milk. He returns with 2 qt. 1 pt. How many pints has he sold?
7. 6s. 4¾d. + 8s. 11¼d.
8. $12 + b + 6 = 20$. What number does b represent?
9. John has 10s. from which he pays 2s. 4d. for a book and 3s. 9d. for a ball. How much has he left?
10. ¼ of a hundred + ¾ of a dozen.

Multiplication and Division

Ex. 3 A

1. 23×9.
2. $222 \div 6$.
3. 3s. 10½d. × 4.
4. 16 gal. 2 qt. ÷ 12.
5. 3 tons 14 cwt. 3 qr. × 3.
6. £18. 12s. 8d. ÷ 8.
7. 43×70.
8. $9500 \div 500$.
9. How many lb. and oz. of sweets shall I require in order to give 29 boys 4 oz. each?
10. I have been given £3. 18s. 4d. to divide among 10 people. How much will each receive?

Ex. 3 B

1. 32×8.
2. $238 \div 7$.
3. 4s. 8d. × 5.
4. 15 yd. 1 ft. 9 in. ÷ 11.
5. 2 tons 16 cwt. 2 qr. × 4.
6. £20. 16s. 3d. ÷ 9.
7. 52×60.
8. $7200 \div 400$.
9. How many lb. and oz. of sweets shall I require in order to give 33 girls 2 oz. each?
10. I have been given £3. 5s. 0d. to divide among 12 people. How much will each receive?

ANSWERS

Ex. 2 A

1. 99.
2. 4s. 2¼d.
3. 6 tons 12 cwt. 0 qr.
4. 914.
5. 3 gal. 3 qt. 1 pt.
6. 1 yd. 2 ft. 6 in.
7. 14s. 6¼d.
8. $a = 5$.
9. 12s. 11d.
10. 102.

Ex. 2 B

1. 116.
2. £1. 1s. 5½d.
3. 24 gal. 0 qt. 0 pt.
4. 814.
5. 2 yd. 1 ft. 3 in.
6. 4 gal. 3 qt. 1 pt.
7. 15s. 4d.
8. $b = 2$.
9. 3s. 11d.
10. 34.

Ex. 3 A

1. 207.
2. 37.
3. 15s. 6d.
4. 1 gal. 1 qt. 1 pt.
5. 11 tons 4 cwt. 1 qr.
6. £2. 6s. 7d.
7. 3010.
8. 19.
9. 7 lb. 4 oz.
10. 7s. 10d.

Ex. 3 B

1. 256.
2. 34.
3. £1. 3s. 4d.
4. 1 yd. 1 ft. 3 in.
5. 11 tons 6 cwt. 0 qr.
6. £2. 6s. 3d.
7. 3120.
8. 18.
9. 4 lb. 2 oz.
10. 5s. 5d.

NOTES AND MEMORANDA

Reduction

1. How many pence are there in 4*s.* 7*d.*?

2. Change 89 pence to shillings and pence.

3. How many gallons are there in 144 pints?

4. Change 76 fourpences to £ *s. d.*

5. Change 17 six-and-eightpences to £ *s. d.*

6. Add together 14 threepences, 9 florins, and 24 three-halfpences and give your answer in £ *s. d.*

7. Write down how many minutes there are from 11.40 a.m. to 2.10 p.m.

8. What is the total number of pounds in 1 ton 1 cwt. 1 qr.?

9. Change 81 inches to yards, feet, inches.

10. What is the total in £ *s. d.* of 23 half-crowns, 15 sixpences, and 20 pence?

Reduction (*continued*)

1. How many pence are there in 3*s.* 9*d.*?

2. Change 101 pence to shillings and pence.

3. Change 9 gal. 3 qt. 1 pt. to pints.

4. How many threepenny pieces can I get for £1. 5*s.* 6*d.*?

5. Change 37 half-crowns to £ *s. d.*

6. Add together 30 pence, 7 three-and-fourpences, and 19 sixpences and give your answer in £ *s. d.*

7. How many hours are there in 210 minutes?

8. What is the total number of inches in 1 yd. 2 ft. 8 in.?

9. How many 2 oz. packets of tea could be made up from ½ cwt.?

10. What is the total in £ *s. d.* of 11 six-and-eightpences, 7 shillings, and 40 halfpence?

ANSWERS

Ex. 4A

1. 55.　　2. 7s. 5d.　　3. 18.　　4. £1. 5s. 4d.　　5. £5. 13s. 4d.

6. £1. 4s. 6d.　　7. 150.　　8. 2380.　　9. 2 yd. 0 ft. 9 in.　　10. £3. 6s. 8d.

Ex. 4B

1. 45.　　2. 8s. 5d.　　3. 79.　　4. 102.　　5. £4. 12s. 6d.

6. £1. 15s. 4d.　　7. 3½.　　8. 68.　　9. 448.　　10. £4. 2s. 0d.

NOTES AND MEMORANDA

Short Methods in Multiplication and Division

Ex. 5A. 1. 237×1000.

2. $750 \div 50$.

3. 39×40.

4. $280 \div 20$.

5. 48×25.

6. $375 \div 25$.

7. 72×125.

8. $3250 \div 125$.

9. 43×101.

10. 67×99.

B. 1. 426×1000.

2. $950 \div 50$.

3. 47×30.

4. $180 \div 60$.

5. 36×25.

6. $525 \div 25$.

7. 56×125.

8. $4750 \div 125$.

9. 51×101.

10. 56×99.

Costs of Dozens

Ex. 6

A. Find the cost of 1 dozen at:

1. $9d$. each.

2. $1s. 1\frac{1}{2}d$. each.

3. $10\frac{1}{4}d$. each.

4. $4\frac{3}{4}d$. each.

Find the cost of $2\frac{1}{2}$ dozens at:

5. $11d$. each.

6. $8\frac{1}{2}d$. each.

Find the cost of 1 at:

7. $6s. 6d$. per doz.

8. $10s. 9d$. per doz.

9. $18s. 6d$. per 2 doz.

10. $14s$. per $3\frac{1}{2}$ doz.

B. Find the cost of 1 dozen at:

1. $7d$. each.

2. $1s. 3\frac{1}{2}d$. each.

3. $8\frac{1}{4}d$. each.

4. $6\frac{3}{4}d$. each.

Find the cost of $2\frac{1}{2}$ dozens at:

5. $10d$. each.

6. $7\frac{1}{2}d$. each.

Find the cost of 1 at:

7. $4s. 6d$. per doz.

8. $9s. 3d$. per doz.

9. $15s. 6d$. per 2 doz.

10. $18s$. per $4\frac{1}{2}$ doz.

ANSWERS

Ex. 5 A

1. 237,000.
2. 15.
3. 1560.
4. 14.
5. 1200.
6. 15.
7. 9000.
8. 26.
9. 4343.
10. 6633.

Ex. 5 B

1. 426,000.
2. 19.
3. 1410.
4. 3.
5. 900.
6. 21.
7. 7000.
8. 38.
9. 5151.
10. 5544.

Ex. 6 A

1. 9s.
2. 13s. 6d.
3. 10s. 3d.
4. 4s. 9d.
5. £1. 7s. 6d.
6. £1. 1s. 3d.
7. $6\frac{1}{2}d$.
8. $10\frac{3}{4}d$.
9. $9\frac{1}{4}d$.
10. 4d.

Ex. 6 B

1. 7s.
2. 15s. 6d.
3. 8s. 3d.
4. 6s. 9d.
5. £1. 5s. 0d.
6. 18s. 9d.
7. $4\frac{1}{2}d$.
8. $9\frac{1}{4}d$.
9. $7\frac{3}{4}d$.
10. 4d.

NOTES AND MEMORANDA

More Costs

Ex. 7

A. Find the cost of:

1. 10 articles at 4d. each.
2. 4 doz. articles at 7½d. each.
3. 43 articles at 2s. 6d. each.
4. 3 doz. articles at 4s. each.
5. 22 articles at 3s. 4d. each.
6. 51 articles at 1s. 11d. each.
7. 16 articles at 3s. 1½d. each.
8. 241 articles at 5½d. each.
9. 478 articles at 7¼d. each.
10. 963 articles at 3¾d. each.

B. Find the cost of:

1. 11 articles at 3d. each.
2. 6 doz. articles at 4½d. each.
3. 52 articles at 2s. each.
4. 2½ doz. articles at 5s. each.
5. 26 articles at 6s. 8d. each.
6. 44 articles at 2s. 1d. each.
7. 24 articles at 1s. 10½d. each.
8. 239 articles at 8½d. each.
9. 482 articles at 6¼d. each.
10. 957 articles at 2¾d. each.

Costs of Weights

Ex. 8

A. What is the cost of 2½ lb. at:

1. 1d. per oz.?
2. 1½d. per oz.?
3. 2½d. per oz.?

At 8d. per lb. find:

4. The cost of 13 oz.
5.   ,,   2 lb. 8 oz.
6.   ,,   5 lb. 9 oz.

At 2s. 8d. per lb. find:

7. The cost of 11 oz.
8.   ,,   1 lb. 14 oz.
9.   ,,   3 lb. 5 oz.
10.   ,,   5 lb. 9 oz.

B. What is the cost of 2½ lb. at:

1. ½d. per oz.?
2. 2d. per oz.?
3. 3d. per oz.?

At 1s. 4d. per lb. find:

4. The cost of 15 oz.
5.   ,,   3 lb. 2 oz.
6.   ,,   4 lb. 7 oz.

At 2s. per lb. find:

7. The cost of 13 oz.
8.   ,,   1 lb. 12 oz.
9.   ,,   4 lb. 3 oz.
10.   ,,   2 lb. 11 oz.

ANSWERS

Ex. 7 A

1. 3s. 4d.   2. £1. 10s. 0d.   3. £5. 7s. 6d.   4. £7. 4s. 0d.   5. £3. 13s. 4d.

6. £4. 17s. 9d.   7. £2. 10s. 0d.   8. £5. 10s. 5½d.   9. £14. 8s. 9½d.   10. £15. 0s. 11¼d.

Ex. 7 B

1. 2s. 9d.   2. £1. 7s. 0d.   3. £5. 4s. 0d.   4. £7. 10s. 0d.   5. £8. 13s. 4d.

6. £4. 11s. 8d.   7. £2. 5s. 0d.   8. £8. 9s. 3½d.   9. £12. 11s. 0½d.   10. £10. 19s. 3¾d.

Ex. 8 A

1. 3s. 4d.   2. 5s.   3. 8s. 4d.   4. 6½d.   5. 1s. 8d.

6. 3s. 8½d.   7. 1s. 10d.   8. 5s.   9. 8s. 10d.   10. 14s. 10d.

Ex. 8 B

1. 1s. 8d.   2. 6s. 8d.   3. 10s.   4. 1s. 3d.   5. 4s. 2d.

6. 5s. 11d.   7. 1s. 7½d.   8. 3s. 6d.   9. 8s. 4½d.   10. 5s. 4½d.

NOTES AND MEMORANDA

The Costs of Scores

Ex. 9

A. Find the cost of a score at:
1. 12s. each.
2. 8s. 6d. each.
3. 15s. 3d. each.
4. 18s. 9d. each.

Find the cost of 1 at:
5. £17 per score.
6. £9. 10s. per score.
7. £14. 15s. per score.
8. £24. 5s. per score.
9. £21. 15s. per 3 score.
10. £20. 12s. 6d. per 2½ score.

B. Find the cost of a score at:
1. 14s. each.
2. 5s. 6d. each.
3. 17s. 9d. each.
4. 13s. 3d. each.

Find the cost of 1 at:
5. £18 per score.
6. £11. 10s. per score.
7. £16. 15s. per score.
8. £19. 5s. per score.
9. £25 per 4 score.
10. £13. 2s. 6d. per 1½ score.

The Costs of Grosses

Ex. 10

A. Find the cost of 1 gross at:
1. 7d. each.
2. 8½d. each.
3. 5¼d. each.
4. 6¾d. each.

Find the cost of 3 gross at:
5. 4½d. each.
6. 10¼d. each.

Find the cost of 1 at:
7. £4. 16s. per gross.
8. £2. 14s. per gross.
9. £2. 5s. per gross.
10. £27. 15s. per 5 gross.

B. Find the cost of 1 gross at:
1. 9d. each.
2. 3½d. each.
3. 6¼d. each.
4. 8¾d. each.

Find the cost of 3 gross at:
5. 7½d. each.
6. 8¾d. each.

Find the cost of 1 at:
7. £5. 8s. per gross.
8. £1. 10s. per gross.
9. £3. 3s. per gross.
10. £18. 12s. per 4 gross.

ANSWERS

Ex. 9 A

1. £12.	2. £8. 10s. 0d.	3. £15. 5s. 0d.	4. £18. 15s. 0d.	5. 17s.
6. 9s. 6d.	7. 14s. 9d.	8. £1. 4s. 3d.	9. 7s. 3d.	10. 8s. 3d.

Ex. 9 B

1. £14.	2. £5. 10s. 0d.	3. £17. 15s. 0d.	4. £13. 5s. 0d.	5. 18s.
6. 11s. 6d.	7. 16s. 9d.	8. 19s. 3d.	9. 6s. 3d.	10. 8s. 9d.

Ex. 10 A

1. £4. 4s. 0d.	2. £5. 2s. 0d.	3. £3. 3s. 0d.	4. £4. 1s. 0d.	5. £8. 2s. 0d.
6. £18. 9s. 0d.	7. 8d.	8. $4\frac{1}{2}d$.	9. $3\frac{3}{4}d$.	10. $9\frac{1}{4}d$.

Ex. 10 B

1. £5. 8s. 0d.	2. £2. 2s. 0d.	3. £3. 15s. 0d.	4. £5. 5s. 0d.	5. £13. 10s. 0d
6. £15. 15s. 0d.	7. 9d.	8. $2\frac{1}{2}d$.	9. $5\frac{1}{4}d$.	10. $7\frac{3}{4}d$.

NOTES AND MEMORANDA

Bills

Ex. 11

A. Find the cost of the following:

1. 9 articles at 8*d*. each.
2. 8 articles at 5¾*d*. each.
3. 42 articles at 1*s*. 3*d*. per doz.
4. 2 doz. articles at 4 for 5*d*.
5. 7 articles at £1. 5*s*. per score.
6. 3½ lb. at 1*s*. 8*d*. per lb.
7. 1½ lb. of butter at 1*s*. 6*d*. per lb. and ¼ lb. of tea at 2*s*. per lb.
8. 2½ lb. of steak at 1*s*. 9*d*. per lb.
9. 3¼ lb. of coffee at 2*s*. 3*d*. per lb.
10. Find the total cost of the items above.

B. Find the cost of the following:

1. 10 articles at 9*d*. each.
2. 16 articles at 2¾*d*. each.
3. 54 articles at 1*s*. 1*d*. per doz.
4. 1½ doz. articles at 3 for 2*d*.
5. 3 articles at £2. 15*s*. per score.
6. 2¼ lb. at 1*s*. 4*d*. per lb.
7. 1¼ lb. of cheese at 8*d*. per lb. and ½ lb. of bacon at 1*s*. 2*d*. per lb.
8. 3½ lb. of beef at 1*s*. 4*d*. per lb.
9. 1¾ lb. of tea at 2*s*. 9*d*. per lb.
10. Find the total cost of the items above.

Change

Ex. 12

A. What amount of change is received in the following cases?

Money proffered	Amount of bill
1. 1*s*.	5½*d*.
2. 2*s*. 6*d*.	1*s*. 4½*d*.
3. 2 florins.	2*s*. 5*d*.
4. 3 half-crowns.	5*s*. 4½*d*.
5. A 10*s*. note.	7*s*. 3*d*.
6. 16*s*.	14*s*. 2½*d*.
7. A £1 note.	18*s*. 2*d*.
8. £2.	£1. 2*s*. 9*d*.
9. £5.	£4. 7*s*. 11*d*.
10. £6. 15*s*.	£6. 13*s*. 7½*d*.

B. What amount of change is received in the following cases?

Money proffered	Amount of bill
1. 1*s*.	8½*d*.
2. 2*s*.	9½*d*.
3. 2 half-crowns.	3*s*. 7*d*.
4. 4 florins.	6*s*. 2½*d*.
5. A 10*s*. note.	8*s*. 8*d*.
6. 17*s*. 6*d*.	15*s*. 3½*d*.
7. A £1 note.	14*s*. 4*d*.
8. £3.	£2. 11*s*. 10*d*.
9. £4. 10*s*.	£4. 3*s*. 5*d*.
10. £5. 15*s*.	£5. 14*s*. 5½*d*.

ANSWERS

Ex. 11A

1. 6*s*.	2. 3*s*. 10*d*.	3. 4*s*. 4½*d*.	4. 2*s*. 6*d*.	5. 8*s*. 9*d*.
6. 5*s*. 10*d*.	7. 2*s*. 9*d*.	8. 4*s*. 4½*d*.	9. 7*s*. 3¾*d*.	10. £2. 5*s*. 8¾*d*.

Ex. 11B

1. 7*s*. 6*d*.	2. 3*s*. 8*d*.	3. 4*s*. 10½*d*.	4. 1*s*.	5. 8*s*. 3*d*.
6. 3*s*.	7. 1*s*. 5*d*.	8. 4*s*. 8*d*.	9. 4*s*. 9¾*d*.	10. £1. 19*s*. 2¼*d*.

Ex. 12A

1. 6½*d*.	2. 1*s*. 1½*d*.	3. 1*s*. 7*d*.	4. 2*s*. 1½*d*.	5. 2*s*. 9*d*.
6. 1*s*. 9½*d*.	7. 1*s*. 10*d*.	8. 17*s*. 3*d*.	9. 12*s*. 1*d*.	10. 1*s*. 4½*d*.

Ex. 12B

1. 3½*d*.	2. 1*s*. 2½*d*.	3. 1*s*. 5*d*.	4. 1*s*. 9½*d*.	5. 1*s*. 4*d*.
6. 2*s*. 2½*d*.	7. 5*s*. 8*d*.	8. 8*s*. 2*d*.	9. 6*s*. 7*d*.	10. 6½*d*.

NOTES AND MEMORANDA

Perimeters and Areas of Rectangles

Ex. 13

A. Find the perimeters of rectangles having the following dimensions:

Length	Breadth
1. 11 in.	6 in.
2. 1 ft. 3 in.	7 in.
3. 10½ in.	8 in.
4. 4 ft. 5½ in.	1 ft. 6 in.
5. 3½ yd.	2 ft.

Find the areas of rectangles having the following dimensions. Give your answer in square inches.

Length	Breadth
6. 9 in.	9 in.
7. 12 in.	7 in.
8. 1 ft. 4 in.	5 in.
9. 11½ in.	9 in.
10. 8½ in.	6½ in.

B. Find the perimeters of rectangles having the following dimensions:

Length	Breadth
1. 10 in.	9 in.
2. 1 ft. 5 in.	4 in.
3. 9½ in.	7 in.
4. 3 ft. 6 in.	2 ft. 4½ in.
5. 2 yd.	2½ ft.

Find the areas of rectangles having the following dimensions. Give your answer in square inches.

Length	Breadth
6. 7 in.	7 in.
7. 11 in.	8 in.
8. 1 ft. 5 in.	6 in.
9. 10½ in.	7 in.
10. 9½ in.	4½ in.

Squares and Square Roots

Ex. 14 A

1. (a) 3^2.   (b) 6^2.
2. (a) 8^2.   (b) 11^2.
3. (a) 20^2.   (b) 50^2.
4. (a) $7^2 - 5^2$.   (b) $3^2 + 2^2$.
5. (a) $\sqrt{25}$.   (b) $\sqrt{49}$.
6. (a) $\sqrt{16}$.   (b) $\sqrt{144}$.
7. (a) $\sqrt{900}$.   (b) $\sqrt{1600}$.
8. (a) $\sqrt{64} - \sqrt{49}$.
   (b) $\sqrt{25} + \sqrt{36}$.
9. The side of a square field is 120 yd. long. What is its area in square yards?
10. What is the perimeter of a square field with an area of 6400 sq. yd.?

Ex. 14 B

1. (a) 4^2.   (b) 5^2.
2. (a) 7^2.   (b) 12^2.
3. (a) 30^2.   (b) 60^2.
4. (a) $8^2 - 6^2$.   (b) $4^2 + 1^2$.
5. (a) $\sqrt{36}$.   (b) $\sqrt{64}$.
6. (a) $\sqrt{81}$.   (b) $\sqrt{121}$.
7. (a) $\sqrt{400}$.   (b) $\sqrt{4900}$.
8. (a) $\sqrt{81} - \sqrt{25}$.
   (b) $\sqrt{16} + \sqrt{144}$.
9. The side of a square field is 110 yd. long. What is its area in square yards?
10. What is the perimeter of a square field with an area of 3600 sq. yd.?

ANSWERS

Ex. 13 A

1. 2 ft. 10 in.	2. 3 ft. 8 in.	3. 3 ft. 1 in.	4. 3 yd. 2 ft. 11 in.	5. 8 yd. 1 ft.
6. 81 sq. in.	7. 84 sq. in.	8. 80 sq. in.	9. 103½ sq. in.	10. 55¼ sq. in.

Ex. 13 B

1. 3 ft. 2 in.	2. 3 ft. 6 in.	3. 2 ft. 9 in.	4. 3 yd. 2 ft. 9 in.	5. 5 yd. 2 ft.
6. 49 sq. in.	7. 88 sq. in.	8. 102 sq. in.	9. 73½ sq. in.	10. 42¾ sq. in.

Ex. 14 A

1. (a) 9.   (b) 36.
2. (a) 64.   (b) 121.
3. (a) 400. (b) 2500.
4. (a) 24.   (b) 13.
5. (a) 5.   (b) 7.
6. (a) 4.   (b) 12.
7. (a) 30.   (b) 40.
8. (a) 1.   (b) 11.
9. 14,400 sq. yd.
10. 320 yd.

Ex. 14 B

1. (a) 16.   (b) 25.
2. (a) 49.   (b) 144.
3. (a) 900. (b) 3600.
4. (a) 28.   (b) 17.
5. (a) 6.   (b) 8.
6. (a) 9.   (b) 11.
7. (a) 20.   (b) 70.
8. (a) 4.   (b) 16.
9. 12,100 sq. yd.
10. 240 yd.

NOTES AND MEMORANDA

The Miscellaneous Arithmetic Exercises
of Daily Life

Ex. 15 A

1. What is the cost of $3\frac{1}{2}$ lb. of bacon at 1s. 2d. per lb.?
2. What change will be received from a 10s. note after paying a bill of 2s. $9\frac{1}{2}d.$?
3. Which is the cheaper and by how much each: (a) 4s. 6d. a dozen or (b) £2. 8s. 0d. per gross?
4. The first $1\frac{1}{2}d.$ ticket a 'bus conductor issues is numbered 0315 and the last one 0479. How many of these tickets has he issued?
5. What is the value of the tickets issued in No. 4?
6. A motor-car covers 266 miles on 7 gal. of petrol. What is the average consumption of the car in miles per gallon?
7. At 1s. 6d. per gal., what would 7 gal. of petrol cost?
8. A milkman starts his round with 10 gal. of milk and returns with $2\frac{1}{2}$ gal. At 3d. per pt., what is the value of his sales?
9. What would be the cost of putting a picture rail round a room 14 ft. by 10 ft. at 2s. per yd.?
10. What will $3\frac{1}{2}$ doz. $1\frac{1}{2}d.$ stamps cost?

The Miscellaneous Arithmetic Exercises
of Daily Life (*continued*)

Ex. 15 B

1. What is the cost of $2\frac{1}{2}$ lb. of cheese at 11d. per lb.?
2. How much change will be received from 2 half-crowns after paying a bill of 3s. $8\frac{1}{2}d.$?
3. Which is the cheaper and by how much each: (a) 2s. 6d. each or (b) £4. 10s. 0d. per 2 score?
4. The first 6d. ticket issued by the pay-box girl at the cinema is numbered 0412 and the last 0537. How many 6d. tickets has she issued?
5. What is the value of the tickets issued in No. 4?
6. A motor-cycle covers 365 miles on 5 gal. of petrol. How many miles per gallon does the machine average?
7. What will 5 gal. cost at 1s. 7d. per gal.?
8. A newspaper boy starts out with 156 twopenny Sunday papers. He returns with 36. What is the value of his sales?
9. What is the cost of covering a room 12 ft. by 12 ft. with linoleum at 2s. 6d. per sq. ft.?
10. How many $1\frac{1}{2}d.$ stamps can I get with 11s. $10\frac{1}{2}d.$?

ANSWERS

      Ex. 15 B

1. 4s. 1d.
2. 7s. 2½d.
3. £2. 8s. 0d. per gross by ½d. each.
4. 165.
5. £1. 0s. 7½d.
6. 38 m.p.g.
7. 10s. 6d.
8. 15s.
9. £1. 12s. 0d.
10. 5s. 3d.

1. 2s. 3½d.
2. 1s. 3½d.
3. £4. 10s. 0d. per 2 score by 3d. each.
4. 126.
5. £3. 3s. 0d.
6. 73 m.p.g.
7. 7s. 11d.
8. £1.
9. £18.
10. 95.

NOTES AND MEMORANDA

Progress Tests

Ex. 16 A

1. $569 + 78$.

2. From £1. 10s. 0d. take 15s. 8d.

3. How many pounds are there in $1\frac{1}{2}$ tons?

4. 35×99.

5. If 3 score pairs of gloves cost £8. 5s., what would be the cost of 1 pair?

6. Find the total cost of 1 dozen articles at $5\frac{1}{2}d$. each and 9 at 6d. each.

7. How much change should I receive from half-a-crown after paying for $1\frac{1}{2}$ lb. of meat at 1s. 4d. per lb.?

8. (a) 90^2.   (b) $\sqrt{144}$.

9. What is the cost of 483 articles at $5\frac{1}{4}d$. each?

10. How much shall I have to pay for 44 articles if each one costs 3s. 4d.?

Progress Tests (*continued*)

Ex. 16 B

1. $632 - 57$.

2. What is the sum of 14s. $7\frac{1}{2}d$. and 9s. 10d.?

3. How many yards are there in $1\frac{1}{2}$ miles?

4. $6750 \div 125$.

5. If 2 gross of exercise books cost £8. 8s., what would be the cost of 1 book?

6. Find the total cost of 1 dozen articles at $3\frac{1}{4}d$. each and 13 at 3d. each.

7. How much change should I receive from a florin after paying for $1\frac{1}{2}$ lb. of cheese at 10d. per lb.?

8. Find (a) the perimeter and (b) the area of a rectangular hall 20 yd. by 15 yd.

9. What is the cost of 958 articles at $7\frac{1}{2}d$. each?

10. What is the value of 23 articles at 1s. 11d. each?

[43]

ANSWERS

Ex. 16 A

1. 647.
2. 14s. 4d.
3. 3360.
4. 3465.
5. 2s. 9d.
6. 10s.
7. 6d.
8. (a) 8100.   (b) 12.
9. £10. 11s. 3¼d.
10. £7. 6s. 8d.

Ex. 16 B

1. 575.
2. £1. 4s. 5½d.
3. 2640.
4. 54.
5. 7d.
6. 6s. 6d.
7. 9d.
8. (a) 70 yd.   (b) 300 sq. yd.
9. £29. 18s. 9d.
10. £2. 4s. 1d.

NOTES AND MEMORANDA

More Progress Tests

Ex. 17 A

1. 635×9.

2. Find the sum of 12s. 10½d. and 7s. 9d.

3. How many 1d. stamps can I buy with 9s. 6d.?

4. Find the cost of 2½ dozen articles at 8½d. each.

5. If exercise books cost £4. 16s. per gross, what will be the price of half-a-dozen?

6. $875 \div 25$.

7. How much change shall I receive from a £1 note after paying for 2¼ lb. of bacon at 1s. 1d. per lb.?

8. What is the cost of 956 articles at 4½d. each?

9. Find (a) the perimeter and (b) the area of a rectangular hall 30 yd. by 16 yd.

10. What is the value of 16 articles at 3s. 1½d. each?

More Progress Tests (*continued*)

Ex. 17 B

1. $712 \div 8$.

2. From £1. 5s. take 11s. 4½d.

3. What is the cost of 81 penny stamps?

4. What will be the cost of 3½ score of ties at 2s. 6d. each?

5. If eggs cost 2s. 9d. per dozen, what will be the price of 3?

6. 73×101.

7. How much change shall I receive from 2 half-crowns after paying for 2½ lb. of butter at 1s. 6d. per lb.?

8. What is the cost of 243 articles at 9½d. each?

9. (a) 11^2.　　(b) $\sqrt{10,000}$.

10. How much shall I have to pay for 53 articles costing 2s. 6d. each?

ANSWERS

Ex. 17 A

1. 5715.
2. £1. 0s. 7½d.
3. 114.
4. £1. 1s. 3d.
5. 4s.
6. 35.
7. 17s. 3½d.
8. £17. 18s. 6d.
9. (a) 92 yd.   (b) 480 sq. yd.
10. £2. 10s. 0d.

Ex. 17 B

1. 89.
2. 13s. 7½d.
3. 6s. 9d.
4. £8. 15s. 0d.
5. 8¼d.
6. 7373.
7. 1s. 3d.
8. £9. 12s. 4½d.
9. (a) 121.   (b) 100.
10. £6. 12s. 6d.

NOTES AND MEMORANDA

Fractions

A. Reduce to Lowest Terms:

1. (a) $\frac{9}{15}$. (b) $\frac{16}{22}$.
2. (a) $\frac{24}{42}$. (b) $\frac{27}{45}$.

Change to Mixed Numbers:

3. (a) $\frac{8}{5}$. (b) $\frac{10}{3}$.
4. (a) $\frac{53}{7}$. (b) $\frac{103}{8}$.

Change to Improper Fractions:

5. (a) $3\frac{2}{7}$. (b) $4\frac{5}{8}$.
6. (a) $2\frac{7}{10}$. (b) $7\frac{9}{11}$.

Simplify:

7. $\frac{2}{3} + \frac{1}{4}$.
8. $\frac{1}{6} - \frac{1}{8}$.
9. $\frac{8}{9} \times \frac{3}{4}$.
10. $\frac{3}{10} \div \frac{1}{5}$.

B. Reduce to Lowest Terms:

1. (a) $\frac{8}{14}$. (b) $\frac{15}{18}$.
2. (a) $\frac{32}{48}$. (b) $\frac{35}{56}$.

Change to Mixed Numbers:

3. (a) $\frac{7}{5}$. (b) $\frac{12}{5}$.
4. (a) $\frac{62}{9}$. (b) $\frac{95}{6}$.

Change to Improper Fractions:

5. (a) $2\frac{5}{8}$. (b) $5\frac{4}{9}$.
6. (a) $3\frac{5}{12}$. (b) $6\frac{8}{10}$.

Simplify:

7. $\frac{3}{4} + \frac{1}{3}$.
8. $\frac{1}{5} - \frac{1}{7}$.
9. $\frac{5}{6} \times \frac{3}{10}$.
10. $\frac{4}{9} \div \frac{1}{3}$.

Fractional Quantities

A. Find the value of:

1. (a) $\frac{1}{4}$ of 1s. (b) $\frac{1}{8}$ of 1s.
2. (a) $\frac{2}{3}$ of 1s. (b) $\frac{1}{6}$ of 1s.
3. (a) $\frac{1}{3}$ of £1. (b) $\frac{5}{8}$ of £1.
4. (a) $\frac{1}{12}$ of £1. (b) £$\frac{51}{480}$.
5. $\frac{4}{9}$ of 5s. 3d.
6. $\frac{3}{4}$ of 1s. $+ \frac{1}{3}$ of 2s. 6d.
7. (a) $\frac{4}{9}$ of 1 yd. (b) $\frac{2}{3}$ of 1 ft.
8. (a) $\frac{3}{10}$ of 1 hr. (b) $\frac{7}{8}$ of 1 gal.
9. (a) $\frac{1}{7}$ of 1 cwt. (b) $\frac{3}{8}$ of 1 ton.
10. (a) $\frac{2}{11}$ of 1 ml. (b) $\frac{5}{16}$ of 2 lb.

B. Find the value of:

1. (a) $\frac{1}{3}$ of 1s. (b) $\frac{5}{8}$ of 1s.
2. (a) $\frac{3}{4}$ of 1s. (b) $\frac{5}{8}$ of 1s.
3. (a) $\frac{1}{8}$ of £1. (b) $\frac{7}{8}$ of £1.
4. (a) $\frac{1}{16}$ of £1. (b) £$\frac{33}{960}$.
5. $\frac{3}{8}$ of 6s. 8d.
6. $\frac{2}{3}$ of 1s. $+ \frac{1}{5}$ of 2s. 6d.
7. (a) $\frac{5}{8}$ of 1 yd. (b) $\frac{3}{4}$ of 1 ft.
8. (a) $\frac{7}{10}$ of 1 hr. (b) $\frac{3}{8}$ of 1 gal.
9. (a) $\frac{1}{8}$ of cwt. (b) $\frac{2}{5}$ of 1 ton.
10. (a) $\frac{3}{8}$ of 1 ml. (b) $\frac{3}{16}$ of 3 lb.

ANSWERS

Ex. 18A

1. (a) $\frac{2}{3}$. (b) $\frac{8}{11}$. 2. (a) $\frac{4}{7}$. (b) $\frac{3}{4}$. 3. (a) $1\frac{3}{8}$. (b) $3\frac{1}{4}$.

4. (a) $7\frac{1}{3}$. (b) $12\frac{7}{8}$. 5. (a) $2\frac{3}{7}$. (b) $3\frac{7}{8}$. 6. (a) $2\frac{7}{10}$. (b) $9\frac{9}{11}$.

7. $1\frac{1}{2}$. 8. $2\frac{1}{4}$. 9. $\frac{2}{3}$. 10. $1\frac{1}{2}$.

Ex. 18B

1. (a) $\frac{4}{7}$. (b) $\frac{5}{8}$. 2. (a) $\frac{3}{4}$. (b) $\frac{5}{8}$. 3. (a) $2\frac{1}{4}$. (b) $2\frac{2}{5}$.

4. (a) $6\frac{5}{8}$. (b) $15\frac{5}{8}$. 5. (a) $1\frac{7}{10}$. (b) $4\frac{9}{10}$. 6. (a) $4\frac{1}{12}$. (b) $9\frac{3}{10}$.

7. $1\frac{1}{12}$. 8. $\frac{2}{35}$. 9. $\frac{1}{4}$. 10. $1\frac{1}{3}$.

Ex. 19A

1. (a) 3d.   (b) 1½d.
2. (a) 8d.   (b) 2d.
3. (a) 6s. 8d.   (b) 12s. 6d.
4. (a) 1s. 8d.   (b) 2s. 1½d.
5. 2s. 4d.
6. 1s. 7d.
7. (a) 1 ft. 4 in. (b) 8 in.
8. (a) 18 min.   (b) 3 qt. 1 pt.
9. (a) 16 lb.   (b) 12 cwt.
10. (a) 320 yd.   (b) 10 oz.

Ex. 19B

1. (a) 4d.   (b) 7½d.
2. (a) 9d.   (b) 10d.
3. (a) 3s. 4d.   (b) 17s. 6d.
4. (a) 1s. 3d.   (b) 8¼d.
5. 2s. 6d.
6. 1s. 2d.
7. (a) 2 ft. 6 in. (b) 9 in.
8. (a) 42 min.   (b) 1 qt. 1 pt.
9. (a) 14 lb.   (b) 8 cwt.
10. (a) 660 yd.   (b) 9 oz.

NOTES AND MEMORANDA

Decimals

A. Change the following vulgar fractions to decimals:

1. (a) $\frac{17}{20}$.　　(b) $\frac{5}{8}$.

2. (a) $\frac{9}{40}$.　　(b) $\frac{16}{25}$.

Change the following decimals to vulgar fractions:

3. (a) ·45.　　(b) ·375.

4. (a) ·06.　　(b) ·0125.

Simplify the following:

5. ·95 + 1·3.

6. 3·7 − 2·9.

7. 2·63 × 30.

8. 5·5 × ·7.

9. 8·55 ÷ 50.

10. 9·68 ÷ ·8.

B. Change the following vulgar fractions to decimals:

1. (a) $\frac{3}{20}$.　　(b) $\frac{7}{8}$.

2. (a) $\frac{11}{40}$.　　(b) $\frac{12}{25}$.

Change the following decimals to vulgar fractions:

3. (a) ·15.　　(b) ·875.

4. (a) ·08.　　(b) ·0625.

Simplify the following:

5. 2·4 + ·76.

6. 4·3 − 2·6.

7. 1·74 × 40.

8. 6·3 × ·8.

9. 6·24 ÷ 40.

10. 6·37 ÷ ·7.

Decimal Quantities

A. Find the value of:

1. ·9 of £1.

2. ·75 of 1s.

3. ·625 of £1.

4. ·65 of £1.

5. ·0875 of £1.

6. ·75 of 1 yd. (in in.).

7. ·6 of 1 ton (in lb.).

8. ·375 of 1 lb. (in oz.).

9. ·85 of 1 hr. (in min.).

10. ·025 of 1 ml. (in yd.).

B. Find the value of:

1. ·6 of £1.

2. ·375 of 1s.

3. ·25 of £1.

4. ·85 of £1.

5. ·0375 of £1.

6. ·25 of 1 gal. (in pt.).

7. ·8 of 1 ml. (in yd.).

8. ·125 of 1 cwt. (in lb.).

9. ·875 of 1 lb. (in oz.).

10. ·025 of 1 ton (in lb.).

ANSWERS

Ex. 20 A

1. (a) ·35. (b) ·625.   2. (a) ·225. (b) ·64.   3. (a) $\frac{9}{20}$. (b) $\frac{3}{8}$.
4. (a) $\frac{3}{50}$. (b) $\frac{1}{80}$.   5. 2·25.   6. ·8.
7. 78·9.   8. 3·85.   9. ·171.   10. 12·1.

Ex. 20 B

1. (a) ·15. (b) ·875.   2. (a) ·275. (b) ·48.   3. (a) $\frac{3}{20}$. (b) $\frac{7}{8}$.
4. (a) $\frac{9}{25}$. (b) $\frac{1}{10}$.   5. 3·16.   6. 1·7.
7. 69·6.   8. 5·04.   9. ·156.   10. 9·1.

Ex. 21 A

1. 18s.   2. 9d.   3. 12s. 6d.   4. 13s.   5. 1s. 9d.
6. 27 in.   7. 1344 lb.   8. 6 oz.   9. 51 min.   10. 44 yd.

Ex. 21 B

1. 12s.   2. 4½d.   3. 5s.   4. 17s.   5. 9d.
6. 2 pt.   7. 1408 yd.   8. 14 lb.   9. 14 oz.   10. 56 lb.

NOTES AND MEMORANDA

Miscellaneous Exercises

Ex. 22 A

1. Add together ·85 and $\frac{2}{5}$ and give your answer as a decimal.

2. What will be the distance covered by a train in 4 hr. if it travels at an average of 52·25 m.p.h.?

3. Each week a young man saves $\frac{2}{5}$ of his wages of £1. 17s. 6d. What amount does he save weekly?

4. 2·96 ÷ ·8.

5. What fraction of 1 yd. is 1 ft. 4 in.?

6. Take 3s. 6d. from ·75 of £1.

7. The sum of ·125 and ·375 of a sum of money is 2s. 3d. What is this sum of money?

8. $\frac{5}{7}$ of 5s. 10d.

9. $\frac{11}{80}$ of £1 + ·9 of £1.

10. If I have travelled $\frac{4}{9}$ of my journey of 180 miles, how many miles have I still to go?

Miscellaneous Exercises (*continued*)

Ex. 22 B

1. Add together $\frac{5}{8}$ and ·7 and give your answer as a decimal.

2. A motor-car covers 306·5 ml. in 5 hr. What is its average speed in m.p.h.?

3. A man working in a Corporation Department is to retire on a pension equal to $\frac{2}{3}$ of his wages. If he earns £3. 7s. 6d. per week, what will be his weekly pension?

4. 37·5 × ·6.

5. What decimal part of 1 gal. is 2 qt. 1 pt.?

6. Add together 12s. 6d. and ·625 of £1.

7. The difference between ·5 and ·75 of a sum of money is 6d. What is the sum of money?

8. $\frac{4}{5}$ of 7s. 6d.

9. $\frac{9}{40}$ of £1 + ·55 of £1.

10. 20 miles of a journey of 140 miles have been covered. What fraction of the journey still remains?

ANSWERS

Ex. 22A

1. 1·25. 2. 209 ml. 3. 15*s*. 4. 3·7. 5. $\frac{1}{5}$.

6 11*s*. 6*d*. 7. 4*s*. 6*d*. 8. 4*s*. 2*d*. 9. £1. 0*s*. 9*d*. 10. 100 ml.

Ex. 22B

1. 1·325. 2. 61·3 m.p.h. 3. £2. 5*s*. 0*d*. 4. 22·5. 5. ·625.

6. £1. 5*s*. 0*d*. 7. 2*s*. 8. 6*s*. 9. 15*s*. 6*d*. 10. $\frac{4}{9}$.

NOTES AND MEMORANDA

A. Find the averages of the following:

1. 6; 8; 10.
2. 4; 7; 11; 14.
3. £1. 3*s*. 5*d*.; £2. 6*s*. 7*d*.
4. 4 hr. 15 min.; 6 hr. 25 min.
5. 1 gal. 3 qt. 1 pt.; 4 gal. 0 qt. 1 pt.
6. 2 yd. 0 ft. 8 in.; 4 yd. 1 ft. 4 in.
7. 5·27 and 3·61.
8. A motorist drives 112·5 ml. in 3 hr. What is his average speed in m.p.h.?
9. How far can a man walk in 8 hr. at 5·125 m.p.h.?
10. At 12 m.p.h., how long will it take a motor-boat to travel 198 miles?

B. Find the averages of the following:

1. 5; 9; 13.
2. 3; 8; 10; 11.
3. £1. 7*s*. 6*d*.; £2. 12*s*. 6*d*.
4. 3 hr. 35 min.; 5 hr. 25 min.
5. 3 gal. 0 qt. 1 pt.; 5 gal. 1 qt. 1 pt.
6. 3 yd. 1 ft. 7 in.; 2 yd. 1 ft. 5 in.
7. 4·16 and 6·24.
8. A man cycles 96·6 ml. in 7 hr. What is his average speed in m.p.h.?
9. How far can a boy cycle in 4 hr. at 13·25 m.p.h.?
10. At 12 m.p.h., how long will it take a trawler to travel 162 miles?

Ex. 24   Approximations and Rough Estimates

A. Find rough answers (in whole numbers) to the following:

1. 4 lb. 15 oz. at 1*s*. 1*d*. per lb.
2. 7 lb. 3 oz. at 5½*d*. per lb.
3. 8 tons 17 cwt. at £5. 1*s*. 10*d*. per ton.
4. 118 × 21.   5. 214 ÷ 71.

Express the following to the nearest shilling:

6. (*a*) 2*s*. 10½*d*.   (*b*) 3*s*. 4½*d*.
7. (*a*) 6*s*. 6*d*.   (*b*) 9*s*. 6½*d*.

Express the following correct to the first decimal place:

8. (*a*) 4·44.   (*b*) 1·08.
9. (*a*) 6·83.   (*b*) 7·75.
10. (*a*) 10·96.   (*b*) 3·03.

B. Find rough answers (in whole numbers) to the following:

1. 6 lb. 4 oz. at 11¼*d*. per lb.
2. 5 lb. 13 oz. at 6¾*d*. per lb.
3. 7 tons 2 cwt. at £3. 18*s*. 9*d*. per ton.
4. 102 × 39.   5. 183 ÷ 62.

Express the following to the nearest shilling:

6. (*a*) 3*s*. 9½*d*.   (*b*) 5*s*. 3½*d*.
7. (*a*) 8*s*. 6*d*.   (*b*) 7*s*. 5½*d*.

Express the following correct to the first decimal place:

8. (*a*) 6·66.   (*b*) 3·14.
9. (*a*) 5·07.   (*b*) 2·25.
10. (*a*) 7·02.   (*b*) 8·98.

ANSWERS

Ex. 23 A

1. 8.   2. 9.   3. £1. 15s. 0d.   4. 5 hr. 20 min.
5. 3 gal.   6. 3 yd. 1 ft. 0 in.   7. 4·44.
8. 37·5 m.p.h.   9. 41 ml.   10. 16½ hr.

Ex. 23 B

1. 9.   2. 8.   3. £2.   4. 4 hr. 30 min.
5. 4 gal. 1 qt. 0 pt.   6. 3 yd.   7. 5·2.
8. 13·8 m.p.h.   9. 53 ml.   10. 13½ hr.

Ex. 24 A

1. 5s.   2. 3s. 6d.   3. £45.   4. 2400.   5. 3.
6. (a) 3s.   (b) 3s.   7. (a) 7s.   (b) 10s.   8. (a) 4·4.   (b) 1·1.
9. (a) 6·8.   (b) 7·8.   10. (a) 11·0.   (b) 3·0.

Ex. 24 B

1. 6s.   2. 3s.   3. £28.   4. 4000.   5. 3.
6. (a) 4s.   (b) 5s.   7. (a) 9s.   (b) 7s.   8. (a) 6·7.   (b) 3·1.
9. (a) 5·1.   (b) 2·3.   10. (a) 7·0.   (b) 9·0.

NOTES AND MEMORANDA

Time-tables

Ex. 25A. The following are the times of departure and arrival of 10 trains running between Birmingham and Leicester. Find the number of hours and minutes taken over each journey:

	Departure	Arrival	Time taken
1.	7.30 a.m.	8.45 a.m.	
2.	8.30 a.m.	9.39 a.m.	
3.	11.48 a.m.	1.50 p.m.	
4.	2.25 p.m.	3.23 p.m.	
5.	3.55 p.m.	5.36 p.m.	
6.	4.55 p.m.	6.3 p.m.	
7.	6.5 p.m.	7.31 p.m.	
8.	7.18 p.m.	8.52 p.m.	
9.	9.30 p.m.	10.42 p.m.	
10.	11.5 p.m.	12.32 a.m.	

Time-tables (*continued*)

Ex. 25B. The following are the times of departure and arrival of 10 trains running between Birmingham and Bristol. Find the number of hours and minutes taken over each journey:

	Departure	Arrival	Time taken
1.	7.30 a.m.	10.0 a.m.	
2.	9.21 a.m.	11.36 a.m.	
3.	10.22 a.m.	12.30 p.m.	
4.	10.45 a.m.	1.16 p.m.	
5.	11.25 a.m.	1.25 p.m.	
6.	1.39 p.m.	3.32 p.m.	
7.	2.0 p.m.	4.43 p.m.	
8.	3.46 p.m.	5.54 p.m.	
9.	3.54 p.m.	6.5 p.m.	
10.	9.10 p.m.	11.30 p.m.	

ANSWERS

Ex. 25 A

1. 1 hr. 15 min.
2. 1 hr. 9 min.
3. 2 hr. 2 min.
4. 58 min.
5. 1 hr. 41 min.
6. 1 hr. 8 min.
7. 1 hr. 26 min.
8. 1 hr. 34 min.
9. 1 hr. 12 min.
10. 1 hr. 27 min.

Ex. 25 B

1. 2 hr. 30 min.
2. 2 hr. 15 min.
3. 2 hr. 8 min.
4. 2 hr. 31 min.
5. 2 hr.
6. 1 hr. 53 min.
7. 2 hr. 43 min.
8. 2 hr. 8 min.
9. 2 hr. 11 min.
10. 2 hr. 20 min.

NOTES AND MEMORANDA

[49*a*]

The Miscellaneous Arithmetic Exercises of Daily Life

Ex. 26 A

1. $3\frac{1}{2}$ lb. of coffee cost 7s. 7d. What is the price per lb.?
2. What change shall I receive from 3 half-crowns after paying a bill of 5s. $10\frac{1}{2}d$.?
3. What is the cost of 1 gal. 2 qt. 1 pt. at 4s. per gal.?
4. A train leaving London at 10.15 a.m. for Birmingham took 2 hr. 20 min. over the journey. What time did it arrive in Birmingham?
5. A boy in charge of a petrol pump found the meter read 087 gal. After the morning sales it read 111 gal. How many gallons had he sold during the morning?
6. At 1s. 6d. per gal., what was the value of his petrol sales?
7. How much shall I have to pay for 117 penny stamps?
8. Which is the cheaper and by what fraction of 1d. each: (a) 4 for 5d. or (b) 6 for 7d.?
9. At 1s. 1d. per lb., how much shall I have to pay for $2\frac{1}{4}$ lb.?
10. During a sale a reduction of $1\frac{1}{2}d$. in the shilling was made in the price of all marked goods. What was the sale price of an article marked at 4s.?

The Miscellaneous Arithmetic Exercises of Daily Life (*continued*)

Ex. 26 B

1. When cheese is 10d. per lb., what will $1\frac{3}{4}$ lb. cost?
2. What change shall I receive from a 10s. note after paying a bill of 7s. $2\frac{1}{2}d$.?
3. What is the cost of 1 yd. 1 ft. 6 in. at 4s. 4d. per yd.?
4. A train leaving Bristol at 11.56 a.m. arrived in London at 2.12 p.m. How many hours and minutes did the journey take?
5. An ice-cream salesman sells 72 twopenny cups during the day. What is the value of his sales?
6. If the ice-cream salesman is allowed a commission of 1d. in the shilling, how much will he get for his sales?
7. How many $1\frac{1}{2}d$. stamps can be bought for 8s. $10\frac{1}{2}d$.?
8. Which is the cheaper and by what fraction of 1d. each: (a) 3 for 2d. or (b) 10d. per doz.?
9. $2\frac{1}{2}$ lb. of tea cost 5s. 10d. What is the price per lb.?
10. If marked prices are reduced by 2s. 6d. in the £ at a sale, what is the sale price of an article marked at £8?

[50]

ANSWERS

Ex. 26 A

1. 2s. 2d.

2. 1s. 7½d.

3. 6s. 6d.

4. 12·35 p.m.

5. 24.

6. £1. 16s. 0d.

7. 9s. 9d.

8. 6 for 7d. by $\frac{1}{12}d$. each.

9. 2s. 5¼d.

10. 3s. 6d.

Ex. 26 B

1. 1s. 5½d.

2. 2s. 9½d.

3. 6s. 6d.

4. 2 hr. 16 min.

5. 12s.

6. 1s.

7. 71.

8. 3 for 2d. by ⅛d. each.

9. 2s. 4d.

10. £7.

NOTES AND MEMORANDA

Triangles, Parallelograms, Circles and Polygons

Ex. 27 A

1. What is the area of a triangle with a base 1 ft. 3 in. and a perpendicular height of 8 in.?
2. A triangle of 60 sq. in. has a base 10 in. long. What is its perpendicular height?
3. Two angles of a triangle measure 40° and 60°. What is the size of the remaining angle?

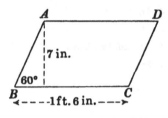

4. What is the area of the parallelogram?
5. What is the size of ∠A?
6. What is the size of ∠D?
7. The radius of a halfpenny is ½ in. What is its circumference? (π = 3·14.)
8. What is the area of a circle with a diameter of 6 in.? (π = 3·14.)
9. The circle of No. 8 just fits into a square. What is the difference in area between the circle and the square?
10. What is an octagon?

Triangles, Parallelograms, Circles and Polygons
(continued)

Ex. 27 B

1. What is the area of a triangle with a base 1 ft. 2 in. and a perpendicular height of 9 in.?
2. A triangle of 48 sq. in. has a perpendicular height of 12 in. What is the length of its base?
3. Two angles of a triangle measure 30° and 50°. What is the size of the remaining angle?

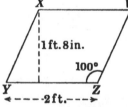

4. What is the area of the parallelogram?
5. What is the size of ∠Y?
6. What is the size of ∠X?
7. The diameter of a circle is 4 in. What is its area? (π = 3·14.)
8. What is the circumference of a circle with a 2 in. radius? (π = 3·14.)
9. The circle of No. 8 just fits inside a square. What is the difference in area between the circle and the square?
10. What is a hexagon?

[51]

ANSWERS

Ex. 27 A

1. 60 sq. in.
2. 1 ft.
3. 80°.
4. 126 sq. in.
5. $\angle A = 120°$.
6. $\angle D = 60°$.
7. 3·14 in.
8. 28·26 sq. in.
9. 7·74 sq. in.
10. An eight-sided plane figure.

Ex. 27 B

1. 63 sq. in.
2. 8 in.
3. 100°.
4. 480 sq. in.
5. $\angle Y = 80°$.
6. $\angle X = 100°$.
7. 12·56 sq. in.
8. 12·56 in.
9. 3·44 sq. in.
10. A six-sided plane figure.

NOTES AND MEMORANDA

Revision Tests

Ex. 28 A

1. Add together 2 tons 11 cwt. 3 qr. and 1 ton 16 cwt. 2 qr.
2. How many half-crowns are there in £6. 12s. 6d.?
3. 65 × 99.
4. Find the cost of 2½ dozen articles at 9d. each?
5. What change shall I receive from a florin after paying for 2½ lb. of liver at 7d. per lb.?
6. What is the perimeter of a square field which has an area of 6400 sq. yd.?
7. $\frac{7}{9} \times \frac{3}{14}$.
8. What is the average weight of 4 packages weighing 7 lb., 9 lb., 10 lb. and 4 lb., respectively?
9. ·792 ÷ ·06.
10. If a train leaving Liverpool at 11.16 a.m. takes 58 min. to travel to Manchester, what will be the time of its arrival in Manchester?

Revision Tests (continued)

Ex. 28 B

1. Take 3 yd. 2 ft. 7 in. from 5 yd. 1 ft. 4 in.
2. Change 53 sixpences to £ s. d.
3. 2250 ÷ 125.
4. Find the cost of 2 score of articles at 4s. 9d. each.
5. How much change shall I receive from half-a-crown after paying for 3½ lb. of cheese at 8d. per lb.?
6. What is the circumference of a circle having a radius of 4 in.? (π = 3·14.)
7. $\frac{3}{4} \div \frac{9}{16}$.
8. An aeroplane travels 864·5 ml. in 7 hr. What is its average speed in miles per hour?
9. 6·75 × ·6.
10. £2. 15s. is shared between Mr Smith and his wife so that the husband gets $\frac{3}{5}$ and Mrs Smith the remainder. How much does each of these receive?

ANSWERS

Ex. 28 A

1. 4 tons 8 cwt. 1 qr.
2. 53.
3. 6435.
4. £1. 2*s*. 6*d*.
5. 6½*d*.
6. 320 yd.
7. ⅛.
8. 7½ lb.
9. 13·2.
10. 12·14 p.m.

Ex. 28 B

1. 1 yd. 1 ft. 9 in.
2. £1. 6*s*. 6*d*
3. 18.
4. £9. 10*s*. 0*d*.
5. 2*d*.
6. 25·12 in.
7. 1⅓.
8. 123·5 m.p.h.
9. 4·05.
10. Mr Smith has £1. 13*s*. 0*d*.
    Mrs Smith has £1. 2*s*. 0*d*.

NOTES AND MEMORANDA

Progress Tests

Ex. 29 A

1. A load of potatoes weighing 1 ton 4 cwt. 3 qr. is to be divided amongst 9 families. What weight of potatoes will each family receive?

2. $925 \div 25$.

3. What will be the cost of 1 article at 11s. 3d. per 3 dozen?

4. What is the total cost of $1\frac{1}{2}$ lb. of mutton at 1s. 9d. per lb. and $\frac{1}{2}$ lb. of steak at 1s. 6d. per lb.?

5. How much shall I have to pay for $2\frac{1}{2}$ gross of articles at 5d. each?

6. What is the area of a triangle with an 11 in. base and a perpendicular height of 9 in.?

7. $\frac{2}{3} + \frac{1}{5}$.

8. From ·625 of £1 take $\frac{1}{3}$ of £1.

9. $9\cdot04 - \cdot375$.

10. An express train leaves Birmingham for London at 10.49 a.m. It arrives in London at 12.17 p.m. How many hours and minutes did it take over the journey?

Progress Tests (*continued*)

Ex. 29 B

1. A bonus of 2s. $10\frac{1}{2}d$. is to be paid out to 8 people. How much money will be required for this purpose?

2. 74×101.

3. What is the cost of 1 article at £11. 10s. 0d. per 2 score?

4. What change shall I receive from 3 florins after paying a bill of 4s. $8\frac{1}{2}d$.?

5. What is the cost of $5\frac{1}{2}$ dozen articles at $2\frac{1}{2}d$. each?

6. Two angles of a triangle measure 30° and 80°. What is the size of the third angle?

7. $\frac{4}{5} - \frac{1}{8}$.

8. From ·75 of £1 take $\frac{2}{3}$ of £1.

9. $2\cdot57 + \cdot068$.

10. A train leaving Wolverhampton for Torquay at 9.34 a.m. takes 5 hr. 18 min. over the journey. What is the time of its arrival in Torquay.

ANSWERS

Ex. 29 A

1. 2 cwt. 3 qr.   2. 37.   3. $3\frac{3}{4}d$.   4. $3s.\ 4\frac{1}{2}d$.   5. £7. 10s. 0d.

6. $49\frac{1}{2}$ sq. in.   7. $1\frac{11}{13}$.   8. 5s. 10d.   9. 8·665.   10. 1 hr. 28 min.

Ex. 29 B

1. £1. 3s. 0d.   2. 7474.   3. 5s. 9d.   4. $1s.\ 3\frac{1}{4}d$.   5. 13s. 9d.

6. 70°.   7. $1\frac{8}{15}$.   8. 1s. 8d.   9. 2·638.   10. 2·52 p.m.

NOTES AND MEMORANDA

More Progress Tests

Ex. 30 A

1. Find the total of 1s. 5d., 2s. 3d. and 4s. 9d.
2. 38 × 40.
3. What is the cost of 2½ score pairs of gloves at 4s. 6d. per pair?
4. What change shall I receive from a £1 note after paying a bill of 13s. 1½d.?
5. What is the value of 32 articles at 1s. 1½d. each?
6. Reduce to their lowest terms (a) $\frac{12}{28}$ and (b) $\frac{18}{63}$.
7. 24·3 × ·7.
8. What is the circumference of a circle with a radius of 5 in.? ($\pi = 3·14$.)
9. $\frac{2}{3}$ of 1s. + ·875 of 1s.
10. At 1s. 4d. per hr., what will a man earn during a morning if he works from 8.45 a.m. until 12.15 p.m.?

More Progress Tests (*continued*)

Ex. 30 B

1. After calling at the grocer's where I spent 4s. 2½d. I find that I have 15s. 9½d. left. With how much did I start out?
2. 850 ÷ 50.
3. At £7. 13s. for 3 gross of articles, what will be the price of 1 of them?
4. What is the total cost of 1½ lb. of cheese at 9d. per lb. and ½ lb. of butter at 1s. 5d. per lb.?
5. Find the average of 35 min., 1 hr. 15 min. and 1 hr. 25 min.
6. ·968 ÷ ·04.
7. $\frac{8}{21} \times \frac{7}{12}$.
8. What is the area of a circle which has a diameter of 2 in.? ($\pi = 3·14$.)
9. Add together $\frac{4}{9}$ of 1 yd. and ·75 of 1 ft. and give your answer in inches.
10. During a sale a reduction of 2d. in the shilling was made on the price of all marked goods. What was the sale price of a hat marked at 5s. 6d.?

ANSWERS

Ex. 30 A

1. 8*s*. 5*d*.
2. 1520.
3. £11. 5*s*. 0*d*.
4. 6*s*. 10½*d*.
5. £1. 16*s*. 0*d*.

6. (*a*) ¾. (*b*) ⅔.
7. 17·01.
8. 31·4 in.
9. 1*s*. 6½*d*.
10. 4*s*. 8*d*.

Ex. 30 B

1. £1.
2. 17.
3. 4¼*d*.
4. 1*s*. 10*d*.
5. 1 hr. 5 min.

6. 24·2.
7. ⅔.
8. 3·14 sq. in.
9. 25 in.
10. 4*s*. 7*d*.

NOTES AND MEMORANDA

Percentages

Ex. 31

A. Change the following to percentages:

1. (a) $\frac{3}{10}$. (b) $\frac{1}{4}$.
2. (a) $\frac{7}{20}$. (b) $\frac{19}{50}$.
3. (a) ·8. (b) ·375.
4. (a) ·65. (b) ·09.

Change the following percentages to vulgar fractions:

5. (a) 30%. (b) 35%.
6. (a) 25%. (b) $87\frac{1}{2}$%.
7. (a) $33\frac{1}{3}$%. (b) 175%.

Change the following percentages to decimals:

8. (a) 19%. (b) 80%.
9. (a) $12\frac{1}{2}$%. (b) 65%.
10. (a) $57\frac{1}{2}$%. (b) $32\frac{1}{2}$%.

B. Change the following to percentages:

1. (a) $\frac{3}{4}$. (b) $\frac{9}{10}$.
2. (a) $\frac{11}{20}$. (b) $\frac{8}{25}$.
3. (a) ·6. (b) ·625.
4. (a) ·35. (b) ·05.

Change the following percentages to vulgar fractions:

5. (a) 70%. (b) 15%.
6. (a) 75%. (b) $62\frac{1}{2}$%.
7. (a) $66\frac{2}{3}$%. (b) 125%.

Change the following percentages to decimals:

8. (a) 13%. (b) 60%.
9. (a) $37\frac{1}{2}$%. (b) 95%.
10. (a) $82\frac{1}{2}$%. (b) $67\frac{1}{2}$%.

Miscellaneous Exercises on Percentages

Ex. 32

A.
1. 50% of 400.
2. $37\frac{1}{2}$% of 72.
3. 60% of £1.
4. 15% of £1.
5. $27\frac{1}{2}$% of £1.
6. 5% of 1 ton (in lb.).
7. $62\frac{1}{2}$% of 1 lb. (in oz.).
8. 75% of 1s. + $12\frac{1}{2}$% of 1s.
9. From 80% of £1 take 25% of 10s.
10. Find the sum of $17\frac{1}{2}$% of £1 and 50% of 9s.

B.
1. 20% of 500.
2. $12\frac{1}{2}$% of 120.
3. 90% of £1.
4. 35% of £1.
5. $22\frac{1}{2}$% of £1.
6. 10% of 1 ml. (in yd.).
7. $87\frac{1}{2}$% of 3 gal. (in pt.).
8. 25% of 1s. + $62\frac{1}{2}$% of 1s.
9. Add together 15% of £1 and 50% of 15s.
10. What is the difference between $2\frac{1}{2}$% of £1 and $12\frac{1}{2}$% of 1s.?

ANSWERS

Ex. 31 A

1. (a) 30 %. (b) 25 %.
2. (a) 35 %. (b) 38 %.
3. (a) 80 %. (b) 37½ %.
4. (a) 65 %. (b) 9 %.
5. (a) $\frac{3}{10}$. (b) $\frac{7}{10}$.
6. (a) $\frac{1}{4}$. (b) $\frac{7}{8}$.
7. (a) $\frac{1}{3}$. (b) $1\frac{3}{4}$.
8. (a) ·19. (b) ·8.
9. (a) ·125. (b) ·65.
10. (a) ·575. (b) ·325.

Ex. 31 B

1. (a) 75 %. (b) 90 %.
2. (a) 55 %. (b) 32 %.
3. (a) 60 %. (b) 62½ %.
4. (a) 35 %. (b) 5 %.
5. (a) $\frac{7}{10}$. (b) $\frac{3}{20}$.
6. (a) $\frac{3}{4}$. (b) $\frac{1}{8}$.
7. (a) $\frac{2}{3}$. (b) $1\frac{1}{4}$.
8. (a) ·13. (b) ·6.
9. (a) ·375. (b) ·95.
10. (a) ·825. (b) ·675.

Ex. 32 A

1. 200. 2. 27. 3. 12s. 4. 3s. 5. 5s. 6d.
6. 112 lb. 7. 10 oz. 8. 10½d. 9. 13s. 6d. 10. 8s.

Ex. 32 B

1. 100. 2. 15. 3. 18s. 4. 7s. 5. 4s. 6d.
6. 176 yd. 7. 21 pt. 8. 10½d. 9. 10s. 6d. 10. 4½d.

NOTES AND MEMORANDA

[55a]

Profit and Loss, and Simple Interest

Ex. 33 A

1. A knife was bought for 2s. and sold for 2s. 6d. What was the gain per cent.?
2. In a case of 200 eggs there were 40 bad ones. What was the percentage of good eggs?
3. I have paid £1 for an article. At what price must I sell it in order to make a profit of 30%?
4. A dealer lost 10% by selling a watch for 4s. 6d. What did the watch cost him?
5. A suite of furniture marked at £32 was reduced 12½%. What was its sale price?

Find the Simple Interest on:

6. £100 for 2 years at 5%.
7. £100 for 3½ years at 4%.
8. £300 for 4 years at 2½%.
9. £250 for 6 years at 3%.
10. £20 for 2½ years at 6%.

Profit and Loss, and Simple Interest (*continued*)

Ex. 33 B

1. An article is bought for 2s. 6d. and sold again for 3s. What is the percentage profit?
2. A boy obtained 270 marks out of a possible 300. What was his percentage loss of marks?
3. I sold an article for 15s. thereby losing 25% on what I had paid for it. What did I pay for it?
4. A dealer buys certain articles at 10s. and intends to sell them at a 20% profit. At what price does he sell them?
5. Fur coats marked at £24 were reduced 37½%. What was the sale price?

Find the Simple Interest on:

6. £100 for 3 years at 4%.
7. £100 for 2½ years at 5%.
8. £400 for 6 years at 3½%.
9. £150 for 5 years at 4%.
10. £70 for 3½ years at 2%.

ANSWERS

NOTES AND MEMORANDA

Commission and Discount

Ex. 34 A

1. How much is a discount of $7\frac{1}{2}\%$ on £5?

2. A discount of 3s. in the £ is allowed. What percentage discount is this?

3. For prompt payments a discount of 10 % is allowed on gas bills. What actual amount do I have to pay for a gas bill of 12s. 6d.?

4. Take 15 % discount off a coat costing £3.

5. Find the amount of a 25 % discount on a bill of £5. 10s.

6. How much in the £ is a commission of (a) 5 %, (b) $1\frac{1}{4}\%$?

7. What will a $2\frac{1}{2}\%$ commission amount to on sales of £12. 10s.?

8. If I am allowed 2s. in the £ on sales, what is my percentage commission?

9. After deducting $12\frac{1}{2}\%$ commission from a sale of £3, what is left?

10. What does a commission of $2\frac{1}{2}\%$ on £320 amount to?

Commission and Discount (*continued*)

Ex. 34 B

1. How much is a discount of $2\frac{1}{2}\%$ on £7?

2. 4s. discount in the £ is allowed. What percentage discount is this?

3. 15 % discount is allowed on electricity bills. What actual amount do I have to pay for an electricity bill of £3?

4. Take $12\frac{1}{2}\%$ discount off a hat costing 16s. 8d.

5. Find the amount after taking a 20 % discount off £2. 10s.

6. How much in the £ is a commission of (a) $2\frac{1}{2}\%$, (b) $6\frac{1}{4}\%$?

7. How much is a $7\frac{1}{2}\%$ commission on sales of £8. 10s.?

8. What percentage commission is 3s. of sales of £1?

9. After deducting 5 % commission from a 15s. sale, what is left?

10. What does a commission of $7\frac{1}{2}\%$ on £200 amount to?

ANSWERS

<table>
<tr><td>

Ex. 34 A

1. 7s. 6d.
2. 15 %.
3. 11s. 3d.
4. £2. 11s. 0d.
5. £1. 7s. 6d.
6. (a) 1s. (b) 3d.
7. 6s. 3d.
8. 10 %.
9. £2. 12s. 6d.
10. £8.

</td><td>

Ex. 34 B

1. 3s. 6d.
2. 20 %.
3. £2. 11s. 0d.
4. 14s. 7d.
5. £2.
6. (a) 6d. (b) 1s. 3d.
7. 12s. 9d.
8. 15 %.
9. 14s. 3d.
10. £15.

</td></tr>
</table>

NOTES AND MEMORANDA

Direct Proportion

Ex. 35 A

1. At 2s. 3d. per dozen, what is the price of 1?
2. If 7 articles cost £1. 15s., what will be the cost of 9?
3. If 11 exercise books cost 4s. 7d., how much shall I have to pay for 3 of them?
4. The cost of 480 articles is £3. 10s. What will be the cost of half-a-dozen such articles?
5. If 2½ lb. of ham cost 5s., what will be the cost of 4 lb.?
6. If buns are 7 for 6d., what will be the cost of 3½ dozen?
7. How much shall I have to pay for 4¼ dozen pencils if 30 of them cost 2s. 6d.?
8. What is the cost of 20 books at 6 for 6s. 3d.?
9. A motor-cyclist can average 45 m.p.h. At this speed how far will he travel in 20 min.?
10. How long will it take the motor-cyclist to travel 9 miles at an average speed of 45 m.p.h.?

Direct Proportion (*continued*)

Ex. 35 B

1. At 3s. 9d. a dozen, what is the price of 1?
2. If 4 articles cost £1. 8s., what will be the cost of 7?
3. How much shall I have to pay for 4 exercise books if 10 of them cost 3s. 4d.?
4. The cost of 240 articles is £2. 15s. What will be the cost of a dozen such articles?
5. If 3½ lb. of steak cost 7s., what will be the cost of 2 lb.?
6. If cakes are 9 for 1s., what will be the cost of 4½ dozen?
7. 20 pencils cost 3s. 4d. How much shall I have to pay for 50 of these pencils?
8. What is the cost of 15 books at 4 for 4s. 8d.?
9. A motorist drives at an average speed of 52 m.p.h. At this speed how far will he travel in 15 min.?
10. How long will it take a motorist to travel 9 miles at an average speed of 54 m.p.h.?

ANSWERS

Ex. 35A

1. $2\frac{1}{4}d.$ 2. £2. 5s. 0d.   3. 1s. 3d.   4. $10\frac{1}{2}d.$ 5. 8s.

6. 3s.   7. 4s. 3d.   8. £1. 0s. 10d.   9. 15 ml.   10. 12 min.

Ex. 35 B

1. $3\frac{3}{4}d.$ 2. £2. 9s. 0d.   3. 1s. 4d.   4. 2s. 9d.   5. 4s.

6. 6s.   7. 8s. 4d.   8. 17s. 6d.   9. 13 ml.   10. 10 min.

NOTES AND MEMORANDA

Inverse Proportion

Ex. 36 A

1. At 4 m.p.h. a man finishes his journey in 5 hr. How long would it have taken him at 5 m.p.h.?

2. How long would rations for 10 men for 5 days last 50 men?

3. 4 men can do a job in 9 days. How long will it take 3 men?

4. 6 boys can do a job in 3 hr. How long will it take 9 boys?

5. John works 3 times as fast as Tom. How long will it take John to do what Tom does in 57 min.?

The cost of carrying 6 tons 3 miles is £1. For the same money find:

6. How far 2 tons should be carried.

7. How far 3 tons should be carried.

8. How far 9 tons should be carried.

9. How far 12 tons should be carried.

10. How far 8 tons should be carried.

Inverse Proportion (*continued*)

Ex. 36 B

1. At 3 m.p.h. a girl finishes her journey in 5 hr. How long would it have taken her at 5 m.p.h.?

2. How long would rations for 12 men for 6 days last 36 men?

3. 5 men can do a job in 8 days. How long will it take 2 men?

4. 8 boys can do a job in 3 hr. How long will it take 12 boys?

5. A man works 5 times as fast as his son. How long will it take the father to do what his son does in 65 min.?

The cost of carrying 9 tons 4 miles is £2. For the same money find:

6. How far 3 tons should be carried.

7. How far 4 tons should be carried.

8. How far 12 tons should be carried.

9. How far 8 tons should be carried.

10. How far 32 tons should be carried.

ANSWERS

Ex. 36 A

1. 4 hr.	2. 1 dy.	3. 12 dy.	4. 2 hr.	5. 19 min.
6. 9 ml.	7. 6 ml.	8. 2 ml.	9. $1\frac{1}{2}$ ml.	10. $2\frac{1}{4}$ ml.

Ex. 36 B

1. 3 hr.	2. 2 dy.	3. 20 dy.	4. 2 hr.	5. 13 min.
6. 12 ml.	7. 9 ml.	8. 3 ml.	9. $4\frac{1}{2}$ ml.	10. $1\frac{1}{4}$ ml.

NOTES AND MEMORANDA

Algebra

A. Simplify:

1. $17x + 8x + 12x$.
2. $43b - 28b$.
3. $5x \times 7y$.
4. $3m \times 14m$.
5. $4t \times 5t \times 6t$.
6. $32y \div 8$.
7. $18x^2 \div 6x$.
8. If $4x = 68$, what is the value of x?
9. If $5x - 3 = 17$, what is the value of x?
10. What is the area of a circle with a radius of $2y$ ft.?

B. Simplify:

1. $11y + 9y + 18y$.
2. $62a - 37a$.
3. $4a \times 9b$.
4. $5n \times 13n$.
5. $3r \times 6r \times 4r$.
6. $48x \div 6$.
7. $24x^2 \div 12x$.
8. If $6x = 90$, what is the value of x?
9. If $8x + 2 = 18$, what is the value of x?
10. What is the circumference of a circle with a radius of $1\frac{1}{2}y$ ft.?

Algebra (*continued*)

A. Simplify:

1. $56y + 78y$.
2. $8a \times 15a$.
3. $2a \times 4b \times 6c$.
4. $198x^2 \div 11x$.
5. $161b \div 7b$.
6. $8a - (3a + 2a)$.
7. $16x - 2(7x - 2x)$.
8. $\dfrac{3y}{5} = 12$.
9. If eggs are y shillings a dozen, what is the cost of 1?
10. At m shillings each, what is the cost in £ of 1 score?

B. Simplify:

1. $69x + 47x$.
2. $7x \times 16x$.
3. $3x \times 4y \times 7z$.
4. $156a^2 \div 12a$.
5. $152y \div 8y$.
6. $9y - (4y - y)$.
7. $21b - 3(3b + 2b)$.
8. $\dfrac{4x}{7} = 20$.
9. If butter is $2x$ shillings a lb., what is the cost of 5 lb.?
10. At n pence each, what is the cost in shillings of 1 dozen?

ANSWERS

Ex. 37 A

1. $37x$.
2. $15b$.
3. $35xy$.
4. $42m^3$.
5. $120t^3$.
6. $4y$.
7. $3x$.
8. $x=17$.
9. $x=4$.
10. $12.56y^2$ sq. ft.

Ex. 37 B

1. $38y$.
2. $25a$.
3. $36ab$.
4. $65n^2$.
5. $72r^3$.
6. $8x$.
7. $2x$.
8. $x=15$.
9. $x=2$.
10. $9.42y$ ft.

Ex. 38 A

1. $134y$.
2. $120a^2$.
3. $48abc$.
4. $18x$.
5. 23.
6. $3a$.
7. $6x$.
8. $y=20$.
9. y pence.
10. £m.

Ex. 38 B

1. $116x$.
2. $112x^2$.
3. $84xyz$.
4. $13a$.
5. 19.
6. $6y$.
7. $6b$.
8. $x=35$.
9. $10x$ shillings.
10. n shillings.

NOTES AND MEMORANDA

The Miscellaneous Arithmetic Exercises of Daily Life

Ex. 39 A

1. What is the cost of $4\frac{1}{4}$ lb. at 2s. 1d. per lb.?
2. Take 2s. $7\frac{1}{2}d$. from 4s. 6d.
3. What percentage reduction is 6s. in the £1?
4. At 1s. 7d. a dozen, what is the cost of 30 eggs?
5. At a sale furs are reduced $37\frac{1}{2}\%$. What is the sale price of a £5 fur?
6. What is the cost of 1 ton 15 cwt. at £4 per ton?
7. A train leaves at 10.29 a.m. and arrives at its destination at 1.57 p.m. How many hours and minutes does the journey take?
8. A sales' reduction is 1s. 6d. in the £. What is the sale price of a £3. 10s. article?
9. What is the cost of 55 articles at 4d. each?
10. At $7\frac{1}{2}\%$, what is the discount on a bill of £2. 10s. 0d.?

The Miscellaneous Arithmetic Exercises of Daily Life (*continued*)

Ex. 39 B

1. How much will $5\frac{1}{2}$ lb. at 2s. 2d. per lb. cost?
2. Take 4s. $9\frac{1}{2}d$. from 6s. 6d.
3. What percentage reduction is $4\frac{1}{2}d$. in the shilling?
4. At 1s. 6d. a dozen, what is the cost of 42 eggs?
5. At a sale coats are reduced $12\frac{1}{2}\%$. What is the sale price of a £4. 4s. coat?
6. What is the cost of 2 tons 10 cwt. at £1. 2s. 6d. per ton?
7. A train leaving at 11.12 a.m. takes $2\frac{3}{4}$ hr. over its journey. At what time does it arrive at its destination?
8. A sales' reduction is 1s. 9d. in the £. What is the sale price of a £4 article?
9. What is the cost of 77 articles at 3d. each?
10. At $2\frac{1}{2}\%$, what is the commission on sales of £14. 10s. 0d.?

ANSWERS

Ex. 39A

1. 8*s.* 10$\frac{1}{4}$*d.*
2. 1*s.* 10$\frac{1}{2}$*d.*
3. 30 %.
4. 3*s.* 11$\frac{1}{2}$*d.*
5. £3. 2*s.* 6*d.*
6. £7.
7. 3 hr. 28 min.
8. £3. 4*s.* 9*d.*
9. 18*s.* 4*d.*
10. 3*s.* 9*d.*

Ex. 39B

1. 11*s.* 11*d.*
2. 1*s.* 8$\frac{1}{2}$*d.*
3. 37$\frac{1}{2}$ %.
4. 5*s.* 3*d.*
5. £3. 13*s.* 6*d.*
6. £2. 16*s.* 3*d.*
7. 1.57 p.m.
8. £3. 13*s.* 0*d.*
9. 19*s.* 3*d.*
10. 7*s.* 3*d.*

NOTES AND MEMORANDA

The Miscellaneous Arithmetic Exercises
of Daily Life (*continued*)

Ex. 40 A

1. How much will 57 articles at 6*d*. each cost?
2. What change shall I receive from £1 after paying a bill of 16*s*. 7*d*.?
3. Find the cost of 2¾ lb. at 2*s*. 2*d*. per lb.
4. At 1*s*. per pint, what will a 5 gal. drum of oil cost?
5. 6 gal. of petrol cost 8*s*. 6*d*. What is the price per gal.?
6. At 5%, what is the commission on sales of £5. 5*s*.
7. I receive discount of 1*s*. 6*d*. on a bill of 10*s*. What percentage discount is this?
8. What is the average weekly wage of a man who earns £2. 12*s*. 6*d*. one week and £3. 2*s*. 6*d*. the next?
9. How many hours and minutes are there from 9.35 a.m. until 3.17 p.m.?
10. What is the cost of 5 gal. 3 qt. at 2*s*. 4*d*. per gal.

The Miscellaneous Arithmetic Exercises
of Daily Life (*continued*)

Ex. 40 B

1. What shall I have to pay for 1½ dozen articles at 2*s*. 6*d*. each?
2. How much change shall I receive from £1 after paying a bill of 12*s*. 4*d*.?
3. Find the cost of 3¼ lb. at 1*s*. 8*d*. per lb.
4. What is the cost of 4½ gal. of petrol at 1*s*. 7*d*. per gal.?
5. If a 5 gal. drum of oil costs £1, what is the price per pt.?
6. At 15%, what is the discount on £6. 10*s*.?
7. I receive 4*s*. 6*d*. commission on a £3 sale. What is my percentage commission?
8. What is the average weekly wage of a woman who earns £1. 17*s*. 6*d*. one week and £1. 12*s*. 6*d*. the next?
9. How many hours and minutes are there from 8.45 a.m. until 3.28 p.m.?
10. What is the cost of 5 yd. 2 ft. at 2*s*. 3*d*. per yd.?

ANSWERS

Ex. 40 A

1. £1. 8s. 6d.
2. 3s. 5d.
3. 5s. 11¼d.
4. £2.
5. 1s. 5d.

6. 5s. 3d.
7. 15 %.
8. £2. 17s. 6d.
9. 5 hr. 42 min.
10. 13s. 5d.

Ex. 40 B

1. £2. 5s. 0d.
2. 7s. 8d.
3. 5s. 5d.
4. 7s. 1½d.
5. 6d.

6. 19s. 6d.
7. 7½ %.
8. £1. 15s. 0d.
9. 6 hr. 43 min.
10. 12s. 9d.

NOTES AND MEMORANDA

Progress Tests

Ex. 41 A

1. $13s.\ 6\frac{1}{2}d. + £1.\ 8s.\ 9d.$

2. What is the cost of 34 articles at $3s.\ 4d.$ each?

3. What will $6\frac{1}{2}$ dozen articles at $4\frac{1}{4}d.$ each cost?

4. If $4\frac{1}{2}$ lb. cost $1s.\ 6d.$, what is the price of 1 lb.?

5. $\frac{15}{16} \times \frac{4}{5}$.

6. $12\frac{1}{2}\%$ of $1s. + \frac{5}{8}$ of $1s. + \cdot25$ of $1s.$

7. If 3 books cost $5s.\ 3d.$, how much shall I have to pay for 4 of them?

8. 25% of a school of 208 pupils are absent. How many of them are present?

9. The perimeter of a square field is 280 yd. What is its area?

10. A dealer buys articles at $4s.\ 6d.$ and sells them at $6s.$ What is the percentage profit on his outlay?

Progress Tests (*continued*)

Ex. 41 B

1. $£2.\ 6s.\ 5d. - 17s.\ 10\frac{1}{2}d.$

2. What will 14 articles at $6s.\ 8d.$ each cost?

3. Find the cost of $7\frac{1}{2}$ dozen articles at $3\frac{1}{2}d.$ each.

4. If $3\frac{1}{2}$ lb. cost $2s.\ 4d.$, what is the price per lb.?

5. $\frac{16}{21} \div \frac{2}{3}$.

6. $\cdot375$ of $1s. + \frac{1}{8}$ of $1s. + 75\%$ of $1s.$

7. If 4 books cost $5s.\ 8d.$, how much shall I have to pay for 6 of them?

8. Out of a school of 200 pupils 40 are absent. What percentage of them are present?

9. A square field has an area of 2500 sq. yd. What is its perimeter?

10. I bought a certain article for $5s.$ and sold it at a 20% loss. How much did I get for it?

ANSWERS

Ex. 41 A

1. £2. 2s. 3½d. 2. £5. 13s. 4d. 3. £1. 9s. 3d. 4. 4d. 5. ¾.

6. 1s. 2½d. 7. 7s. 8. 156. 9. 4900 sq. yd. 10. 33⅓ %.

Ex. 41 B

1. £1. 8s. 6½d. 2. £4. 13s. 4d. 3. £1. 6s. 3d. 4. 8d. 5. 1½.

6. 1s. 3½d. 7. 8s. 6d. 8. 80 %. 9. 200 yd. 10. 4s.

NOTES AND MEMORANDA

Final Tests

Ex. 42 A

1. £5. 16s. 3d. × 6.

2. At £3. 10s. per 2 score, what is the cost of 1?

3. Find the cost of 1½ gross of articles at 6d. each.

4. What change should I receive from 7s. 6d. after paying a bill of 5s. 11d.?

5. 9·76 ÷ ·08.

6. How long would rations for 8 men for 9 days last 18 men?

7. During a sale there was a reduction of 22½ % on marked prices. What was the sale price of an article marked at £3?

8. 30 % of £1 + ·025 of £1 + ⅛ of £1.

9. If $2x/3 = 20$, what is the value of x?

10. What is the Simple Interest on £200 for 5½ years at 3 %?

Final Tests (*continued*)

Ex. 42 B

1. £8. 13s. 3d. ÷ 7.

2. What is the cost of 4½ dozen articles at 2s. 6d. each?

3. 3 gross of articles cost £9. 9s. What is the cost of 1?

4. How much should I have to pay for 2 lb. 15 oz. at 1s. 4d. per lb.?

5. 3·65 × 1·1.

6. It costs £5 to take a load of 6 tons 4 miles. For the same charge how far would a load of 8 tons be taken?

7. How much is a 12½ % commission on sales of £6. 10s. 0d.?

8. 40 % of £1 + ⅓ of £1 + ·075 of £1.

9. If $4x - 2 = 10$, what is the value of x?

10. What is the Simple Interest on £400 for 2½ years at 5 %?

[64]

ANSWERS

Ex. 42 A

1. £34. 17s. 6d.   2. 1s. 9d.   3. £5. 8s. 0d.   4. 1s. 7d.   5. 122.

6. 4 dy.   7. £2. 6s. 6d.   8. 9s. 10d.   9. $x=30$.   10. £33.

Ex. 42 B

1. £1. 4s. 9d.   2. £6. 15s. 0d.   3. $5\frac{1}{4}$.   4. 3s. 11d.   5. 4·015.

6. 3 ml.   7. 16s. 3d.   8. 16s. 2d.   9. $x=3$.   10. £50.

NOTES AND MEMORANDA

PART THREE

Miscellaneous Exercises

Ex. 1 A

1. Take £3. 15s. 3½d. from £5. 19s. 2d.

2. How many are there in 1½ gross?

3. 6875 ÷ 125.

4. What is the cost of 957 articles at 11¼d. each?

5. Share 17s. 6d. between a boy and a girl so that the boy has twice as much as the girl.

6. What is the cost of 40 articles at 3s. 10½d. each?

7. What change shall I receive from a £1 note after paying for 4½ lb. at 1s. 7d. per lb.?

8. How many days are there from the 1st of March until September 30th inclusive?

9. At 8 for 5d., what will be the cost of 1 gross?

10. Find the total cost of 2¾ lb. at 1s. 4d. per lb. and 1½ lb. at 1s. 8d. per lb.

Miscellaneous Exercises (*continued*)

Ex. 1 B

1. Add together £2. 17s. 6d. and £3. 8s. 10d.

2. Take 4½ dozen from 3½ score.

3. 282 × 50.

4. How much shall I have to pay for 484 articles at 10½d. each?

5. Divide £1. 2s. 6d. between a man and his son so that the father receives five times as much as the boy.

6. What is the cost of 60 articles at 2s. 2d. each?

7. How much change shall I receive from a £1 note after paying for 6½ lb. at 1s. 5d. per lb.?

8. How many days are there from the 1st of July until December 31st inclusive?

9. What is the cost of 101 articles at 3s. 4d. each?

10. What is the total cost of 2½ lb. at 1s. 5d. per lb. and 2¾ lb. at 2s. per lb.?

ANSWERS

Ex. 1 A

1. £2. 3s. 10½d.
2. 216.
3. 55.
4. £45. 17s. 1½d.
5. Boy has 11s. 8d. Girl has 5s. 10d.
6. £7. 15s. 0d.
7. 12s. 10½d.
8. 214.
9. 7s. 6d.
10. 6s. 2d.

Ex. 1 B

1. £6. 6s. 4d.
2. 16.
3. 14,100.
4. £21. 3s. 6d.
5. Father has 18s. 9d. Son has 3s. 9d.
6. £6. 10s. 0d.
7. 10s. 9½d.
8. 184.
9. £16. 16s. 8d.
10. 9s. 0½d.

NOTES AND MEMORANDA

Revision Tests in the Four Rules

Ex. 2 A

1. $5896 + 207 + 658$.

2. £5. 9s. $3\frac{1}{2}d. \times 7$.

3. From £9. 12s. $8\frac{1}{2}d$. take £7. 14s. $6\frac{3}{4}d$.

4. 13 yd. 2 ft. 9 in. $\div 6$.

5. $\frac{1}{2}$ gross $+ 1\frac{3}{4}$ score $+ \frac{1}{2}$ dozen.

6. The milkman sets out with $12\frac{1}{2}$ gal. of milk. How much has he left if he sells 10 gal. 3 qt. 1 pt.?

7. What change is received from a £1 note after paying for 2 articles costing 6s. $11\frac{1}{2}d$. and 3s. 5d. respectively?

8. Find the total profit made on 30 articles bought for 4s. 6d. each and sold at 6s. each.

9. How much money is required to pay the wages of £3. 12s. 8d. to each of 11 men?

10. What is the cost of 1 if 9 cost £2. 15s. $10\frac{1}{2}d.$?

Revision Tests in the Four Rules (*continued*)

Ex. 2 B

1. $7345 - 5967$.

2. 13 tons 7 cwt. 3 qr. $\div 9$.

3. Find the sum of £2. 9s. $10\frac{1}{2}d.$, £7. 6s. $4\frac{3}{4}d$. and £5. 10s. $3\frac{1}{2}d$.

4. 5 gal. 3 qt. 1 pt. $\times 8$.

5. $\frac{1}{4}$ gross $+ \frac{1}{2}$ score $+ 1\frac{3}{4}$ dozen.

6. From a roll of cloth $10\frac{1}{2}$ yd. long a tailor cuts a length 3 yd. 2 ft. 8 in. What length of cloth remains on the roll?

7. After paying for 2 articles costing 5s. $9\frac{1}{2}d$. and 3s. 8d., how much change shall I have from a £1 note?

8. What is the total profit made on 54 articles bought for 1s. $10\frac{1}{2}d$. each and sold for 2s. $4\frac{1}{2}d$. each?

9. What total amount is required to pay 12 men their wages of £3. 11s. 4d. each?

10. What is the cost of 1 if 7 cost £2. 11s. $7\frac{1}{2}d.$?

ANSWERS

Ex. 2 A

1. 6761.
2. £38. 5s. 0½d.
3. £1. 18s. 1¾d.
4. 2 yd. 0 ft. 11½ in.
5. 113.
6. 1 gal. 2 qt. 1 pt.
7. 9s. 7½d.
8. £2. 5s. 0d.
9. £39. 19s. 4d.
10. 6s. 2½d.

Ex. 2 B

1. 1378.
2. 1 ton 9 cwt. 3 qr.
3. £15. 6s. 6¾d.
4. 47 gal. 0 qt. 0 pt.
5. 67.
6. 6 yd. 1 ft. 10 in.
7. 10s. 6½d.
8. £1. 7s. 0d.
9. £42. 16s. 0d.
10. 7s. 4½d.

NOTES AND MEMORANDA

Reduction

Ex. 3 A. Bring:

1. 158 pence to *s. d.*
2. 68 in. to yd. etc.
3. 58 pt. to gal. etc.
4. 93 qr. to tons, etc.
5. 29 oz. to lb. etc.

Change:

6. 15*s.* 4*d.* to pence.
7. £10. 17*s.* 6*d.* to half-crowns.
8. 16*s.* 10½*d.* to three-half-pences.
9. 5 yd. 2 ft. 11 in. to inches.
10. How many minutes are there from 11.36 a.m. to 2.25 p.m.?

Ex. 3 B. Bring:

1. 171 pence to *s. d.*
2. 59 in. to yd. etc.
3. 67 pt. to gal. etc.
4. 102 qr. to tons, etc.
5. 31 oz. to lb. etc.

Change:

6. 14*s.* 9*d.* to pence.
7. £9. 12*s.* 6*d.* to half-crowns.
8. 18*s.* 7½*d.* to three-half-pences.
9. 9 gal. 3 qt. 1 pt. to half-pints.
10. How many minutes are there from 10.45 a.m. to 1.34 p.m.?

Short Methods

Ex. 4 A

1. $376 \times 200.$
2. $2250 \div 50.$
3. $464 \times 125.$
4. $99 \times 83.$

Find the cost of:

5. 44 articles at 3*s.* 4*d.* each.
6. 85 „ at 6*d.* each.
7. 238 „ at 1*s.* 1½*d.* each.
8. 18 „ at 4*s.* 1½*d.* each.
9. 2 lb. 13 oz. at 2*s.* per lb.
10. What is the cost of 1½ lb. at 2½*d.* per oz.?

Ex. 4 B

1. $254 \times 300.$
2. $268 \times 25.$
3. $7625 \div 125.$
4. $101 \times 76.$

Find the cost of:

5. 38 articles at 6*s.* 8*d.* each.
6. 53 „ at 3*d.* each.
7. 482 „ at 1*s.* 3½*d.* each.
8. 21 „ 2*s.* 10½*d.* each.
9. 3 lb. 15 oz. at 8*d.* per lb.
10. What is the cost of 2½ lb. at 1½*d.* per oz.?

ANSWERS

Ex. 3 A

1. 13s. 2d.   2. 1 yd. 2 ft. 8 in.   3. 7 gal. 1 qt. 0 pt.   4. 1 ton 3 cwt. 1 qr.
5. 1 lb. 13 oz.   6. 184.   7. 87.   8. 135.   9. 215.   10. 169.

Ex. 3 B

1. 14s. 3d.   2. 1 yd. 1 ft. 11 in.   3. 8 gal. 1 qt. 1 pt.   4. 1 ton 5 cwt. 2 qr.
5. 1 lb. 15 oz.   6. 177.   7. 77.   8. 149.   9. 158.   10. 169.

Ex. 4 A

1. 75,200.   2. 45.   3. 58,000.   4. 8217.   5. £7. 6s. 8d.
6. £2. 2s. 6d.   7. £13. 7s. 9d.   8. £3. 14s. 3d.   9. 5s. 7½d.   10. 5s.

Ex. 4 B

1. 76,200.   2. 6700.   3. 61.   4. 7676.   5. £12. 13s. 4d.
6. 13s. 3d.   7. £31. 2s. 7d.   8. £3. 0s. 4½d.   9. 2s. 7½d.   10. 5s.

NOTES AND MEMORANDA

Costs of Dozens, Scores and Grosses

Ex. 5

A. Find the cost of 2 dozen at:

1. (*a*) 7*d*. ea. (*b*) 8¾*d*. ea.
2. (*a*) 11¼*d*. ea. (*b*) 1*s*. 5½*d*. ea.

Find the cost of 2½ score at:

3. (*a*) 4*s*. ea. (*b*) 7*s*. 6*d*. ea.
4. (*a*) 2*s*. 3*d*. ea. (*b*) 5*s*. 9*d*. ea.

Find the cost of 3 gross at:

5. (*a*) 4*d*. ea. (*b*) 6½*d*. ea.
6. (*a*) 5¼*d*. ea. (*b*) 9¾*d*. ea.

Find the cost of 1 at:

7. 15*s*. 6*d*. per dozen.
8. £9. 5*s*. per score.
9. £3. 6*s*. per gross.
10. £1. 19*s*. per gross.

B. Find the cost of 3 dozen at:

1. (*a*) 5*d*. ea. (*b*) 9¼*d*. ea.
2. (*a*) 10¾*d*. ea. (*b*) 1*s*. 2½*d*. ea.

Find the cost of 1½ score at:

3. (*a*) 6*s*. ea. (*b*) 5*s*. 6*d*. ea.
4. (*a*) 4*s*. 3*d*. ea. (*b*) 3*s*. 9*d*. ea.

Find the cost of 3 gross at:

5. (*a*) 5*d*. ea. (*b*) 7½*d*. ea.
6. (*a*) 4¼*d*. ea. (*b*) 8¾*d*. ea.

Find the cost of 1 at:

7. 16*s*. 6*d*. per dozen.
8. £10. 15*s*. per score.
9. £2. 14*s*. per gross.
10. £1. 13*s*. per gross.

Costs of Dozens, Scores and Grosses (*continued*)

Ex. 6

A. Find the cost of 3½ dozen at:

1. 10½*d*. ea.
2. 1*s*. 3¼*d*. ea.

Find the cost of 6 score at:

3. 3*s*. 6*d*. ea.
4. 4*s*. 9*d*. ea.

Find the cost of 2½ gross at:

5. 6½*d*. ea.
6. 7¼*d*. ea.

Find the cost of 1 at:

7. £1. 6*s*. 3*d*. per 2½ dozen.
8. £15. 15*s*. per 3 score.
9. £5. 2*s*. per 2 gross.
10. £6. 15*s*. per 1½ gross.

B. Find the cost of 2½ dozen at:

1. 11½*d*. ea.
2. 1*s*. 5¾*d*. ea.

Find the cost of 8 score at:

3. 2*s*. 6*d*. ea.
4. 3*s*. 3*d*. ea.

Find the cost of 1½ gross at:

5. 9½*d*. ea.
6. 11¾*d*. ea.

Find the cost of 1 at:

7. £2. 0*s*. 3*d*. per 3½ dozen.
8. £13. 10*s*. per 2 score.
9. £6. 15*s*. per 3 gross.
10. £8. 5*s*. per 2½ gross.

ANSWERS

Ex. 5A

1. (a) 14s. (b) 17s. 6d.
2. (a) £1. 2s. 6d. (b) £1. 15s. 0d.
3. (a) £10. (b) £18. 15s. 0d.
4. (a) £5. 12s. 6d. (b) £14. 7s. 6d.
5. (a) £7. 4s. 0d. (b) £11. 14s. 0d.
6. (a) £9. 9s. 0d. (b) £17. 11s. 0d.
7. 1s. 3½d.
8. 9s. 3d.
9. 5¼d.
10. 3¼d.

Ex. 5B

1. (a) 15s. (b) £1. 7s. 9d.
2. (a) £1. 12s. 3d. (b) £2. 3s. 6d.
3. (a) £9. (b) £8. 5s. 0d.
4. (a) £6. 7s. 6d. (b) £5. 12s. 6d.
5. (a) £9. (b) £13. 10s. 0d.
6. (a) £7. 13s. 0d. (b) £15. 15s. 0d.
7. 1s. 4½d.
8. 10s. 9d.
9. 4½d.
10. 2¾d.

Ex. 6A

1. £1. 16s. 9d. 2. £2. 13s. 4½d. 3. £21. 4. £28. 10s. 0d. 5. £9. 15s. 0d.
6. £10. 17s. 6d. 7. 10½d. 8. 5s. 3d. 9. 4¼d. 10. 7½d.

Ex. 6B

1. £1. 8s. 9d. 2. £2. 4s. 4½d. 3. £20. 4. £26. 5. £8. 11s. 0d.
6. £10. 11s. 6d. 7. 11½d. 8. 6s. 9d. 9. 3¾d. 10. 5½d.

NOTES AND MEMORANDA

Miscellaneous Exercises

Ex. 7 A

1. Find the sum of £29. 12s. 6½d., £5. 8s. 2¼d. and £3. 7s. 5½d.
2. 14 gal. 2 qt. 1 pt. ÷ 9.
3. What is the cost of 4½ dozen articles at 7½d. each?
4. 568 × 125.
5. How much change shall I receive from 10s. after paying for 5½ lb. at 1s. 9d. per lb.?
6. Find the cost of 478 articles at 4¼d. each.
7. Bring 5 gal. 3 qt. 1 pt. to half-pints.
8. 27 × 102.
9. At £11. 5s. 0d. per 2½ gross, what is the price of 1?
10. Find the total cost of 2 lb. 11 oz. at 1s. 4d. per lb. and 1¼ lb. at 1½d. per oz.?

Miscellaneous Exercises (*continued*)

Ex. 7 B

1. From £32. 8s. 6½d. take £17. 14s. 8¾d.
2. 7 yd. 2 ft. 9 in. × 7.
3. Find the cost of 3½ score of articles at 4s. 9d. each.
4. 5375 ÷ 125.
5. What change shall I receive from £1 after paying for 4½ lb. at 1s. 5d. per lb.?
6. How many three-halfpences are there in 17s. 7½d.?
7. Find the cost of 32 articles at 1s. 10½d. each.
8. 34 × 98.
9. At £7. 7s. 0d. per 3½ gross, what is the price of 1?
10. What is the total cost of 1 lb. 13 oz. at 2s. 8d. per lb. and 1¾ lb. at ½d. per oz.?

ANSWERS

Ex. 7 A

1. £38. 8s. 2$\frac{1}{4}d$.   2. 1 gal. 2 qt. 1 pt.   3. £1. 13s. 9d.   4. 71,000.   5. 4$\frac{1}{4}d$.

6. £8. 9s. 3$\frac{1}{4}d$.   7. 94.   8. 2754.   9. 7$\frac{1}{2}d$.   10. 6s. 1d.

Ex. 7 B

1. £14. 13s. 9$\frac{3}{4}d$.   2. 55 yd. 1 ft. 3 in.   3. £16. 12s. 6d.   4. 43.   5. 13s. 7$\frac{1}{2}d$.

6. 141.   7. £3.   8. 3332.   9. 3$\frac{1}{4}d$.   10. 6s.

NOTES AND MEMORANDA

Square Roots and Areas. Circles

Ex. 8A

1. (a) 70^2.   (b) $\sqrt{14400}$.

2. Find the area of a square room which has a perimeter of 42 ft.

3. What is the area of a rectangle 1 ft. $3\frac{1}{2}$ in. by 8 in.?

4. A triangle has a 7 in. base and a perpendicular height of 9 in. What is its area?

5. What is the area of a parallelogram with the same dimensions as the triangle of No. 4?

6. (a) $7^2 - 4^2$.   (b) $\sqrt{81} + \sqrt{121}$.

7. Find the circumference of a circle with radius $1\frac{1}{2}$ in. ($\pi = 3 \cdot 14$.)

8. What is the area of a circle with diameter 6 in.? ($\pi = 3 \cdot 14$.)

Find the missing dimensions in the following:

9. Area of rectangle = 45 sq. in.; length = 10 in.; width = ?.

10. Area of triangle = 29 sq. in.; base = 8 in.; height = ?.

Square Roots and Areas. Circles (*continued*)

Ex. 8B

1. (a) 90^2.   (b) $\sqrt{12100}$.

2. What is the perimeter of a square field which has an area of 6400 sq. yd.?

3. Find the area of a rectangle 1 ft. $7\frac{1}{2}$ in. by 9 in.

4. What is the area of a triangle with an 11 in. base and a perpendicular height of 5 in.?

5. Find the area of a parallelogram with the same dimensions as the triangle of No. 4.

6. (a) $9^2 + 5^2$.   (b) $\sqrt{144} - \sqrt{64}$.

7. Find the circumference of a circle with radius $2\frac{1}{2}$ in. ($\pi = 3 \cdot 14$.)

8. What is the area of a circle with diameter 4 in.? ($\pi = 3 \cdot 14$.)

Find the missing dimensions in the following:

9. Area of rectangle = 76 sq. in.; width = 8 in.; length = ?.

10. Area of triangle = 19 sq. in.; height = 4 in.; base = ?.

[71]

ANSWERS

NOTES AND MEMORANDA

[71a]

Arithmetic in Daily Life

Ex. 9 A

1. Find the cost of 5½ lb. at 1s. 3d. per lb.
2. What change is received from 5s. after paying for 3¼ lb. at 1s. 4d. per lb.?
3. At 6d. per sq. ft., what is the cost of painting a floor 12 ft. by 10¼ ft.?
4. What is earned by a man working 45 hr. at 1s. 4d. per hr.?
5. A man works from 7.45 a.m. till 12.30 p.m. and from 2 p.m. till 4.45 p.m. daily. How many hours does he work from Monday morning till Friday evening?

N.B. *No. of units of electricity used by lamps, etc.* $= \dfrac{wattage \times hr.\ used}{1000}$.

Find the number of electricity units used by:

6. A 60-watt lamp burning for 25 hr.
7. A 100-watt lamp burning for 125 hr.
8. An 80-watt wireless set used for 75 hr.
9. A 3500-watt electric fire alight for 11 hr.
10. At 2½d. a unit, what is the cost of burning six 25-watt lamps for 10 hr.?

Arithmetic in Daily Life (*continued*)

Ex. 9 B

1. What is the cost of 4½ lb. at 1s. 7d. per lb.?
2. How much change is received from 10s. after paying for 2¾ lb. at 2s. per lb.?
3. At 1s. 3d. per sq. ft., what is the cost of covering a floor 12 ft. by 8½ ft.?
4. How many hours does a man work from Monday a.m. till Friday p.m. if daily he works from 8.15 a.m. till 12.45 p.m. and 1.30 p.m. till 5.15 p.m.?
5. What is earned by a man working 44 hr. at 1s. 3d. per hr.?
   N.B. *See N.B. in Ex.* 9 A.

Find the number of electricity units used by:

6. A 40-watt lamp burning for 125 hr.
7. A 100-watt lamp burning for 35 hr.
8. A 60-watt wireless set used for 75 hr.
9. A 2500-watt electric fire alight for 12 hr.
10. At 3d. a unit, what is the cost of burning eight 50-watt lamps for 15 hr.?

ANSWERS

Ex. 9A

1. 6s. 10$\frac{1}{2}$d.
2. 8d.
3. £3. 3s. 0d.
4. £3.
5. 37$\frac{1}{2}$ hr.
6. 1$\frac{1}{2}$.
7. 12$\frac{1}{2}$.
8. 6.
9. 38$\frac{1}{2}$.
10. 3$\frac{3}{4}$d.

Ex. 9B

1. 7s. 1$\frac{1}{2}$d.
2. 4s. 6d.
3. £6. 7s. 6d.
4. 41$\frac{1}{4}$ hr.
5. £2. 15s. 0d.
6. 5.
7. 3$\frac{1}{2}$.
8. 4$\frac{1}{2}$.
9. 30.
10. 1s. 6d.

NOTES AND MEMORANDA

Progress Tests

Ex. 10 A

1. Take 5 tons 15 cwt. 3 qr. from 19 tons 8 cwt. 2 qr.

2. 37 gal. 3 qt. 1 pt. × 11.

3. How many 3d. articles can I buy with £2. 12s. 9d.?

4. Find the cost of 2 lb. 7 oz. at 2s. 8d. per lb.

5. What is the cost of 2½ score at 8s. 9d. each?

6. (a) $9^2 + 11^2$.   (b) $\sqrt{121} - \sqrt{36}$.

7. Find what change should be given from 10s. after paying for 3¼ lb. of bacon at 1s. 3d. per lb.

8. What money is earned by a man working 43 hr. at 1s. 6d. per hr.?

9. How many hours and minutes are there altogether from 7.50 a.m. till 12.25 p.m. and 1.45 p.m. till 5.30 p.m.?

10. At 3d. a unit, what is the cost of having a 2500-watt electric fire alight for 12 hr.?

Progress Tests (*continued*)

Ex. 10 B

1. Add together 7 yd. 2 ft. 8 in. and 6 yd. 2 ft. 11 in.

2. £386. 10s. 6d. ÷ 12.

3. 270 × 80.

4. What is the cost of 3 lb. 13 oz. at 1½d. an oz.?

5. Find the cost of 1½ gross at 10½d. each.

6. What is the area of a square field with a perimeter of 480 yd.?

7. How much change should be received from 5s. after paying for 2¼ lb. of butter at 1s. 5d. per lb.?

8. What are the wages of a man working 49 hr. at 10d. per hr.?

9. How many hours are worked in a day by a man who labours from 7.45 a.m. till 12.30 p.m. and from 1.50 p.m. till 5.20 p.m.?

10. At 2d. a unit, find the cost of using an 80-watt wireless set for 50 hr.

[73]

ANSWERS

Ex. 10 A

1. 13 tons 12 cwt. 3 qr.
2. 416 gal. 2 qt. 1 pt.
3. 211.
4. 6*s*. 6*d*.
5. £21. 17*s*. 6*d*.
6. (*a*) 202.   (*b*) 5.
7. 5*s*. 7½*d*.
8. £3. 4*s*. 6*d*.
9. 8 hr. 20 min.
10. 7*s*. 6*d*.

Ex. 10 B

1. 14 yd. 2 ft. 7 in.
2. £32. 4*s*. 2½*d*.
3. 21,600.
4. 7*s*. 7½*d*.
5. £9. 9*s*. 0*d*.
6. 14,400 sq. yd.
7. 1*s*. 5½*d*.
8. £2. 0*s*. 10*d*.
9. 8¼ hr.
10. 8*d*.

NOTES AND MEMORANDA

More Progress Tests

Ex. 11 A

1. £18. 16s. 3¼d. + £6. 9s. 7½d. + £8. 5s. 6½d.

2. 16 yd. 2 ft. 3 in. ÷ 9.

3. 736 × 125.

4. What is the cost of 4¼ dozen articles at 7½d. each?

5. Find the cost of 1 at £6. 7s. 6d. per 2¼ gross.

6. What change will be received from 3 half-crowns after paying for 5¾ lb. at 1s. 2d. per lb.?

7. What is the perimeter of a square field which has an area of 3600 sq. yd.?

8. What is earned by a man working 48 hr. at 1s. 3d. per hr.?

9. How many units of electricity are consumed by eight 60-watt lamps each burning for 25 hr.?

10. At 3½d. a unit, what is the cost of the electricity consumed?

More Progress Tests (*continued*)

Ex. 11 B

1. Take £2. 17s. 9½d. from £5. 14s. 10¼d.

2. 29 gal. 3 qt. 1 pt. × 8.

3. How many 3s. 4d. are there in £29. 10s. 0d.?

4. Find the cost of 1 at £19. 5s. 0d. per 3½ score.

5. What is the cost of 5¾ dozen articles at 4½d. each?

6. What change will be received from £1 after paying for 4¼ lb. at 1s. 8d. per lb.?

7. (a) $12^2 - 7^2$.   (b) $\sqrt{81} + \sqrt{64}$.

8. How much is earned in 47 hr. at 1s. 6d. per hr.?

9. How many units of electricity are consumed by 125 80-watt lamps burning for 6 hr.?

10. At 2½d. a unit, what is the cost of the electricity consumed?

[74]

ANSWERS

Ex. 11 A

1. £33. 11s. 5$\frac{1}{4}$d.
2. 1 yd. 2 ft. 7 in.
3. 92,000.
4. £1. 11s. 10$\frac{1}{2}$d.
5. 4$\frac{1}{4}$d.
6. 9$\frac{1}{4}$d.
7. 240 yd.
8. £3.
9. 12.
10. 3s. 6d.

Ex. 11 B

1. £2. 17s. 0$\frac{3}{4}$d.
2. 239 gal.
3. 177.
4. 5s. 6d.
5. £1. 5s. 10$\frac{1}{2}$d.
6. 12s. 11d.
7. (a) 95. (b) 17.
8. £3. 10s. 6d.
9. 60.
10. 12s. 6d.

NOTES AND MEMORANDA

Fractions

Ex. 12

A. Reduce to Lowest Terms:

1. (a) $\frac{18}{27}$.　　(b) $\frac{15}{45}$.

2. (a) $\frac{64}{96}$.　　(b) $\frac{132}{165}$.

Change to Mixed Numbers:

3. (a) $\frac{18}{7}$.　　(b) $\frac{47}{3}$.

4. (a) $\frac{79}{5}$.　　(b) $\frac{135}{12}$.

Change to Improper Fractions:

5. (a) $4\frac{5}{6}$.　　(b) $9\frac{7}{8}$.

6. (a) $6\frac{5}{12}$.　　(b) $12\frac{6}{7}$.

Simplify:

7. $\frac{1}{4} + \frac{5}{6}$.

8. $\frac{5}{7} - \frac{3}{8}$.

9. $\frac{9}{16} \times \frac{8}{27}$.

10. $\frac{3}{4} \div \frac{15}{32}$.

B. Reduce to Lowest Terms:

1. (a) $\frac{16}{56}$.　　(b) $\frac{14}{49}$.

2. (a) $\frac{54}{144}$.　　(b) $\frac{108}{216}$.

Change to Mixed Numbers:

3. (a) $\frac{22}{9}$.　　(b) $\frac{53}{4}$.

4. (a) $\frac{71}{3}$.　　(b) $\frac{152}{11}$.

Change to Improper Fractions:

5. (a) $3\frac{4}{7}$.　　(b) $8\frac{5}{9}$.

6. (a) $7\frac{6}{11}$.　　(b) $11\frac{5}{8}$.

Simplify:

7. $\frac{1}{3} + \frac{4}{7}$.

8. $\frac{4}{5} - \frac{3}{4}$.

9. $\frac{14}{15} \times \frac{5}{21}$.

10. $\frac{2}{3} \div \frac{16}{27}$.

Fractional Quantities

Ex. 13

A. Find the value of:

1. (a) $\frac{3}{4}$ of 1s.　　(b) $\frac{5}{6}$ of 1s.

2. (a) $\frac{2}{3}$ of £1.　　(b) $\frac{7}{15}$ of £1.

3. (a) $\frac{9}{10}$ of £1.　　(b) £$\frac{87}{960}$.

4. $\frac{7}{11}$ of 8s. 3d.

5. $1\frac{2}{3}$ of £1. 7s. 6d.

6. $\frac{5}{6}$ of £3. 10s.

7. (a) $\frac{3}{4}$ of 1 yd.　　(b) $\frac{5}{6}$ of 1 ft.

8. (a) $\frac{4}{5}$ of 1 ton.　　(b) $\frac{3}{4}$ of 1 gal.

9. (a) $\frac{5}{7}$ of 1 qr.　　(b) $\frac{9}{20}$ of 1 hr.

10. (a) $\frac{1}{8}$ of 1 ml.　　(b) $\frac{5}{6}$ of 1 lb.

B. Find the value of:

1. (a) $\frac{2}{3}$ of 1s.　　(b) $\frac{3}{8}$ of 1s.

2. (a) $\frac{5}{6}$ of £1.　　(b) $\frac{9}{16}$ of £1.

3. (a) $\frac{7}{8}$ of £1.　　(b) £$\frac{69}{480}$.

4. $\frac{5}{9}$ of 7s. 6d.

5. $1\frac{3}{4}$ of £1. 12s.

6. $\frac{7}{8}$ of £2. 10s.

7. (a) $\frac{7}{9}$ of 1 yd.　　(b) $\frac{3}{4}$ of 1 ft.

8. (a) $\frac{3}{4}$ of 1 ton.　　(b) $\frac{5}{8}$ of 1 gal.

9. (a) $\frac{3}{7}$ of 1 cwt.　　(b) $\frac{4}{15}$ of 1 hr.

10. (a) $\frac{1}{11}$ of 1 ml.　　(b) $\frac{7}{8}$ of 1 lb.

ANSWERS

Ex. 12 A

1. (a) $\frac{3}{8}$. (b) $\frac{1}{3}$.
2. (a) $\frac{4}{5}$. (b) $\frac{5}{8}$.
3. (a) $2\frac{4}{7}$. (b) $15\frac{3}{4}$.
4. (a) $15\frac{1}{4}$. (b) $11\frac{1}{4}$.
5. (a) $\frac{28}{63}$. (b) $\frac{10}{81}$.
6. (a) $\frac{17}{15}$. (b) $\frac{90}{71}$.
7. $1\frac{1}{13}$.   8. $\frac{18}{35}$.
9. $\frac{1}{4}$.   10. $1\frac{5}{8}$.

Ex. 13 A

1. (a) $9d$. (b) $7\frac{1}{2}d$.
2. (a) $13s. 4d.$ (b) $9s. 4d.$
3. (a) $18s.$ (b) $1s. 9\frac{3}{4}d.$
4. $5s. 3d.$
5. £2. $5s. 10d.$
6. £2. $3s. 9d.$
7. (a) 2 ft. 3 in. (b) 10 in.
8. (a) 16 cwt. (b) 3 qt.
9. (a) 20 lb. (b) 27 min.
10. (a) 220 yd. (b) 10 oz.

Ex. 12 B

1. (a) $\frac{4}{7}$. (b) $\frac{3}{7}$.
2. (a) $\frac{3}{8}$. (b) $\frac{1}{2}$.
3. (a) $2\frac{4}{9}$. (b) $13\frac{1}{4}$.
4. (a) $23\frac{3}{4}$. (b) $13\frac{9}{11}$.
5. (a) $\frac{35}{54}$. (b) $\frac{7}{57}$.
6. (a) $\frac{34}{17}$. (b) $\frac{58}{85}$.
7. $\frac{15}{17}$.   8. $\frac{1}{20}$.
9. $\frac{3}{5}$.   10. $1\frac{1}{4}$.

Ex. 13 B

1. (a) $8d$. (b) $4\frac{1}{2}d$.
2. (a) $16s. 8d.$ (b) $11s. 3d.$
3. (a) $17s. 6d.$ (b) $2s. 10\frac{1}{4}d$
4. $4s. 2d.$
5. £2. $16s. 0d.$
6. £2. $3s. 9d.$
7. (a) 2 ft. 4 in. (b) 9 in.
8. (a) 15 cwt. (b) 2 qt. 1 pt.
9. (a) 48 lb. (b) 16 min.
10. (a) 160 yd. (b) 14 oz.

NOTES AND MEMORANDA

Decimals

Ex. 14

A. Change to decimals:

1. (a) $\frac{3}{4}$.   (b) $\frac{17}{20}$.
2. (a) $\frac{11}{40}$.   (b) $\frac{18}{25}$.

Change to Vulgar Fractions:

3. (a) ·65.   (b) ·625.
4. (a) ·07.   (b) ·025.

Simplify:

5. $1·08 + ·795$.
6. $3·06 - 1·78$.
7. $·087 \times 600$.
8. $7·5 \times 1·1$.
9. $·964 \div ·08$.
10. $7·06 \div 50$.

B. Change to decimals:

1. (a) $\frac{3}{8}$.   (b) $\frac{13}{20}$.
2. (a) $\frac{19}{40}$.   (b) $\frac{14}{25}$.

Change to Vulgar Fractions:

3. (a) ·85.   (b) ·125.
4. (a) ·05.   (b) ·075.

Simplify:

5. $2·67 + ·084$.
6. $2·54 - 1·86$.
7. $·069 \times 800$.
8. $8·4 \times 1·2$.
9. $·783 \div ·06$.
10. $6·28 \div 50$.

Decimal Quantities

Ex. 15

A. Find the value of:

1. ·625 of 4s.
2. ·35 of £1.
3. ·75 of £1. 10s.
4. ·025 of £1.
5. £·8 + £1·0125.
6. ·875 of $2\frac{1}{2}$ lb. (in oz.).
7. ·375 of 5 gal. (in pt.).
8. ·65 of 1 hr. (in min.).
9. ·075 of 1 ton (in lb.).
10. ·3125 of £1.

B. Find the value of:

1. ·375 of 3s.
2. ·45 of £1.
3. ·25 of £3. 10s.
4. ·075 of £1.
5. £·3 + £2·0875.
6. ·125 of $4\frac{1}{2}$ lb. (in oz.).
7. ·625 of 3 gal. (in pt.).
8. ·85 of 1 hr. (in min.).
9. ·075 of 1 ml. (in yd.).
10. ·1875 of £1.

[76]

ANSWERS

1. (a) ·75.   (b) ·85.
2. (a) ·275.   (b) ·72.
3. (a) $1\frac{3}{10}$.   (b) $\frac{1}{8}$.
4. (a) $\frac{7}{100}$.   (b) $\frac{1}{40}$.
5. 1·875.   6. 1·28.   7. 52·2.
8. 8·25.   9. 12·05.   10. ·1412.

Ex. 14B

1. (a) ·375.   (b) ·65.
2. (a) ·475.   (b) ·56.
3. (a) $1\frac{7}{10}$.   (b) $\frac{1}{4}$.
4. (a) $\frac{1}{40}$.   (b) $\frac{9}{40}$.
5. 2·754.   6. ·68.   7. 55·2.
8. 10·08.   9. 13·05.   10. ·1256·

Ex. 15A

1. 2s. 6d.   2. 7s.   3. £1. 2s. 6d.   4. 6d.   5. £1. 16s. 3d.
6. 35 oz.   7. 15 pt.   8. 39 min.   9. 168 lb.   10. 6s. 3d.

Ex. 15B

1. 1s. 1½d.   2. 9s.   3. 17s. 6d.   4. 1s. 6d.   5. £2. 7s. 9d.
6. 9 oz.   7. 15 pt.   8. 51 min.   9. 132 yd.   10. 3s. 9d.

NOTES AND MEMORANDA

Miscellaneous Exercises

Ex. 16 A

1. Add together $\frac{17}{20}$ and ·25 and give your answer as a decimal.

2. From £·625 take 5s. 9d.

3. What decimal part of 1 yd. is 2 ft. 3 in.?

4. A boy saves $\frac{4}{5}$ of his weekly pocket-money, which is 1s. 6d. At this rate how long will he be saving £1?

5. Express 3s. 9d. as a decimal of £1.

6. Change £·4375 to shillings and pence.

7. $\frac{4}{5}$ of 3s. 4d. + ·0375 of £1.

8. $3·08 \div ·011$.

9. $\frac{14}{27} \times \frac{18}{35}$.

10. An aeroplane has covered ·625 of its journey of 2240 miles. How many miles has it still to go?

Miscellaneous Exercises (*continued*)

Ex. 16 B

1. Add together ·625 and $\frac{3}{4}$ and give your answer as a decimal.

2. Take 6s. 8d. from £·75.

3. What decimal part of 1 gal. is 2 qt. 1 pt.?

4. Each week a boy saves ·625 of his 1s. pocket-money. At this rate how long will he be saving £1?

5. Express 6s. 3d. as a decimal of £1.

6. Change £·0625 to shillings and pence.

7. ·0875 of £1 + $\frac{3}{7}$ of 2s. 11d.

8. $9·05 \times 1·2$.

9. $\frac{7}{10} \div \frac{14}{25}$.

10. A motorist has covered $\frac{3}{10}$ of his journey of 1760 miles. How many miles has he still to go?

[77]

ANSWERS

NOTES AND MEMORANDA

Volumes

Ex. 17 A. Find the volume of rectangular solids having the following dimensions:

1. Length = 7 in.   Breadth = 4 in.   Height = 9 in.
2.   „   = 8 in.   „   = 3 in.   „   = 6 in.
3.   „   = 12 in.   „   = 9 in.   „   = 10 in.
4.   „   = 1 ft. 6 in.   „   = 5 in.   „   = 8 in.
5. How many cubic in. are there in 3 cubic ft.?
6. Change 399 cubic yd. to cubic ft.
7. What is the volume of a cylinder of length 10 in. and radius 3 in.?
8. How many cubic in. are there in a cone with a diameter of 2 in. and height of 6 in.?
9. Change ·375 cubic ft. to cubic in.
10. A rectangular solid with a volume of 648 cubic in. has a length of 8 in. and a height of 9 in. What is its breadth?

Volumes (*continued*)

Ex. 17 B. Find the volume of rectangular solids having the following dimensions:

1. Length = 8 in.   Breadth = 5 in.   Height = 6 in.
2.   „   = 9 in.   „   = 4 in.   „   = 8 in.
3.   „   = 10 in.   „   = 7 in.   „   = 12 in.
4.   „   = 1 ft. 3 in.   „   = 8 in.   „   = 9 in.
5. How many cubic in. are there in 2 cubic ft.?
6. Change 501 cubic yd. to cubic ft.
7. What is the volume of a cylinder of length 5 in. and radius 2 in.?
8. How many cubic in. are there in a pyramid with a base 4 in. by 7 in. and a height of 9 in.?
9. Change ·625 cubic ft. to cubic in.
10. A rectangular solid has a volume of 819 cubic in., a height of 9 in. and a breadth of 7 in. What is its length?

ANSWERS

Ex. 17A

1. 252 cu. in.   2. 144 cu. in.   3. 1080 cu. in.   4. 720 cu. in.   5. 5184 cu. in.

6. 10,773 cu. ft.   7. 282·6 cu. in.   8. 6·28 cu. in.   9. 648 cu. in.   10. 9 in.

Ex. 17B

1. 240 cu. in.   2. 288 cu. in.   3. 840 cu. in.   4. 1080 cu. in.   5. 3456 cu. in.

6. 13,527 cu. ft.   7. 62·8 cu. in.   8. 84 cu. in.   9. 1080 cu. in.   10. 13 in.

NOTES AND MEMORANDA

More Short Methods and Costs

Ex. 18 A

1. 46×600.
2. 368×25.
3. $9375 \div 125$.
4. 102×57.

Find the cost of:

5. 29 articles at 1s. 8d. each.
6. 67 ,, at 3d. each.
7. 54 ,, at 3s. 10½d. each.
8. 477 ,, at 1s. 2½d. each.
9. 2 lb. 9 oz. at 2s. 8d. per lb.
10. 1 lb. 14 oz. at 1¾d. per oz.

Ex. 18 B

1. 73×400.
2. $3750 \div 50$.
3. 432×125.
4. 98×47.

Find the cost of:

5. 35 articles at 1s. 3d. each.
6. 56 ,, at 4d. each.
7. 78 ,, at 2s. 1½d. each.
8. 243 ,, at 1s. 4½d. each.
9. 3 lb. 6 oz. at 2s. per lb.
10. 1 lb. 12 oz. at 2¼d. per oz.

More Short Methods and Costs (*continued*)

Ex. 19

A. Find the cost of:

1. 3 dozen at 2¾d. each.
2. 2½ dozen at 4¼d. each.
3. 4 score at 6s. 9d. each.
4. 3½ score at 7s. 3d. each.
5. 4 gross at 5½d. each.
6. 2½ gross at 8¼d. each.
7. 1 at £1. 4s. 9d. per dozen.
8. 1 at 15s. 9d. per 3½ dozen.
9. 1 at £15. 12s. 6d. per 2½ score.
10. 1 at £7. 13s. per 1½ gross.

B. Find the cost of:

1. 4 dozen at 1¾d. each.
2. 3½ dozen at 3¼d. each.
3. 5 score at 5s. 9d. each.
4. 2½ score at 8s. 3d. each.
5. 3 gross at 7½d. each.
6. 3½ gross at 4½d. each.
7. 1 at £1. 6s. 3d. per dozen.
8. 1 at 19s. 3d. per 3½ dozen.
9. 1 at £11. 17s. 6d. per 2½ score.
10. 1 at £8. 11s. per 1½ gross.

ANSWERS

Ex. 18 A

1. 27,600.
2. 9200.
3. 75.
4. 5814.
5. £2. 8s. 4d.
6. 16s. 9d.
7. £10. 9s. 3d.
8. £28. 16s. 4½d.
9. 6s. 10d.
10. 4s. 4½d.

Ex. 18 B

1. 29,200.
2. 75.
3. 54,000.
4. 4606.
5. £2. 3s. 9d.
6. 18s. 8d.
7. £8. 5s. 9d.
8. £16. 14s. 1½d.
9. 6s. 9d.
10. 5s. 3d.

Ex. 19 A

1. 8s. 3d.
2. 10s. 7½d.
3. £27.
4. £25. 7s. 6d.
5. £13. 4s. 0d.
6. £12. 7s. 6d.
7. 2s. 0¾d.
8. 4½d.
9. 6s. 3d.
10. 8½d.

Ex. 19 B

1. 7s.
2. 11s. 4½d.
3. £28. 15s. 0d.
4. £20. 12s. 6d.
5. £13. 10s. 0d.
6. £9. 9s. 0d.
7. 2s. 2¼d.
8. 5½d.
9. 4s. 9d.
10. 9½d.

NOTES AND MEMORANDA

Time Sheets

Ex. 20 A. The following are the daily hours of work of various men. Find the total number of hours and minutes worked each day:

	Morning Period	Afternoon Period	Total Time
1.	7.0 a.m.–12.30 p.m.	1.30 p.m.–5.0 p.m.	
2.	7.45 a.m.–12.45 p.m.	1.45 p.m.–5.15 p.m.	
3.	8.15 a.m.–12.30 p.m.	1.15 p.m.–5.30 p.m.	
4.	7.30 a.m.–12.15 p.m.	1.30 p.m.–5.15 p.m.	
5.	8.0 a.m.–12.45 p.m.	2.0 p.m.–5.45 p.m.	
6.	7.45 a.m.– 1.15 p.m.	1.45 p.m.–5.30 p.m.	
7.	7.15 a.m.–12.30 p.m.	1.15 p.m.–4.45 p.m.	
8.	8.30 a.m.– 1.15 p.m.	2.0 p.m.–6.30 p.m.	
9.	7.20 a.m.–12.30 p.m.	1.40 p.m.–5.15 p.m.	
10.	8.30 a.m.– 1.15 p.m.	2.0 p.m.–6.20 p.m.	

Time Sheets (*continued*)

Ex. 20 B. The following are the daily hours of work of various men. Find the total number of hours and minutes worked each day:

	Morning Period	Afternoon Period	Total Time
1.	8.30 a.m.– 1.0 p.m.	2.0 p.m.–5.30 p.m.	
2.	7.15 a.m.–12.30 p.m.	1.45 p.m.–5.0 p.m.	
3.	8.0 a.m. –12.15 p.m.	1.30 p.m.–5.45 p.m.	
4.	7.30 a.m.–12.45 p.m.	1.15 p.m.–5.45 p.m.	
5.	8.15 a.m.– 1.0 p.m.	2.0 p.m.–6.15 p.m.	
6.	7.45 a.m.–12.30 p.m.	1.30 p.m.–5.30 p.m.	
7.	7.0 a.m.–12.45 p.m.	1.30 p.m.–4.45 p.m.	
8.	7.45 a.m.– 1.30 p.m.	2.15 p.m.–6.0 p.m.	
9.	7.40 a.m.–12.20 p.m.	1.30 p.m.–5.20 p.m.	
10.	8.20 a.m.– 1.15 p.m.	2.0 p.m.–5.45 p.m.	

ANSWERS

Ex. 20 A

1. 9 hr. 0 min.
2. 8 hr. 30 min.
3. 8 hr. 30 min.
4. 8 hr. 30 min.
5. 8 hr. 30 min.
6. 9 hr. 15 min.
7. 8 hr. 45 min.
8. 9 hr. 15 min.
9. 8 hr. 45 min.
10. 9 hr. 5 min.

Ex. 20 B

1. 8 hr. 0 min.
2. 8 hr. 30 min.
3. 8 hr. 30 min.
4. 9 hr. 45 min.
5. 9 hr. 0 min.
6. 8 hr. 45 min.
7. 9 hr. 0 min.
8. 9 hr. 30 min.
9. 8 hr. 30 min.
10. 8 hr. 40 min.

NOTES AND MEMORANDA

Arithmetic in Daily Life

Ex. 21 A

1. What is the cost of 5½ dozen articles at 2*s*. 6*d*. each?
2. After paying for 3½ lb. at 1*s*. 5*d*. per lb., what change is received from 10*s*.?
3. Mr Jones works from 8.30 a.m. till 1 p.m. and from 2 p.m. till 5.15 p.m. How many hours and minutes is this?
4. What wages will be received for this work at 1*s*. 4*d*. per hr.?
5. Find the cost of 1½ gross of articles at 5¾*d*. each.
6. How many units of electricity will be used up by 24 60-watt lamps burning for 25 hr.?
7. What is the cost of this electricity at 2¾*d*. a unit?
8. Find the cost of 1 ton 15 cwt. 2 qr. at £2 per ton.
9. What is the hire purchase price of a wireless set at 15*s*. down and 24 instalments of 7*s*. 6*d*.?
10. Find the total cost of the fares for 6 adults and 2 children if adults' tickets are 7*s*. 6*d*. and the children's ⅔ of 7*s*. 6*d*.

Arithmetic in Daily Life (*continued*)

Ex. 21 B

1. Find the cost of 6½ dozen articles at 2*s*. each.
2. What change is received from £1 after paying for 7 lb. at 1*s*. 1½*d*. per lb.?
3. Mr Brown works from 8.15 a.m. till 12.30 p.m. and from 1.45 p.m. till 5.30 p.m. How many hours and minutes is this?
4. What wages will he receive for this work at 1*s*. 3*d*. per hr.?
5. What is the cost of 4¼ dozen articles at 9½*d*. each?
6. Find the number of units of electricity used up by 18 25-watt lamps burning for 30 hr.
7. What is the cost of this electricity at 3½*d*. a unit?
8. Find the cost of 1 ton 10 cwt. 1 qr. at £4 per ton.
9. What is the hire purchase price of a bicycle at 17*s*. 6*d*. down and 15 weekly instalments of 5*s*.?
10. Find the total cost of the fares for 3 adults and 4 children if adults' tickets are 4*s*. 6*d*. and the children's ⅔ of 4*s*. 6*d*.

ANSWERS

Ex. 21 A

1. £8. 5s. 0d.   2. 5s. 0½d.   3. 7 hr. 45 min.   4. 10s. 4d.   5. £5. 3s. 6d.

6. 36.   7. 8s. 3d.   8. £3. 11s. 0d.   9. £9. 15s. 0d.   10. £2. 15s. 0d.

Ex. 21 B

1. £7. 16s. 0d.   2. 12s. 1¼d.   3. 8 hr. 0 min.   4. 10s.   5. £2. 0s. 4½d.

6. 13½.   7. 3s. 11¼d.   8. £6. 1s. 0d.   9. £4. 12s. 6d.   10. £1. 5s. 6d.

NOTES AND MEMORANDA

Progress Tests

Ex. 22 A

1. 5 gal. 3 qt. 1 pt. + 7 gal. 2 qt. 0 pt. + 3 gal. 3 qt. 1 pt.

2. How many $1\frac{1}{2}d$. stamps can I buy with $16s.$ $4\frac{1}{2}d.$?

3. At $2s.$ $8\frac{1}{2}d.$ per $3\frac{1}{4}$ dozen, what is the price of 1 dozen?

4. Find the area of a triangle with a base of 9 in. and a height of $3\frac{1}{2}$ in.

5. $\frac{4}{7} - \frac{5}{9}$.

6. $\cdot 048 \times 125$.

7. $\frac{4}{5}$ of $4s.$ $7d.$ + $\cdot 4125$ of £1.

8. What is the volume of a wall 9 ft. long, 6 ft. high and 6 in. thick?

9. How many hours and minutes are there from 9.37 a.m. until 5.12 p.m.?

10. At $2\frac{1}{2}d.$ a unit, how long can I allow my 80-watt wireless set to be on so that it does not use more than $10d.$ worth of electricity?

Progress Tests (*continued*)

Ex. 22 B

1. 2 yd. 1 ft. 8 in. + 6 yd. 2 ft. 7 in. + 4 yd. 1 ft. 3 in.

2. 496×125.

3. How much shall I have to pay for $6\frac{1}{2}$ score at $1s.$ $9d.$ each?

4. What is the area of a triangle with a base of $5\frac{1}{2}$ in. and a height of 11 in.?

5. $\frac{3}{8} + \frac{5}{11}$.

6. $\cdot 0425 \div \cdot 25$.

7. $\frac{2}{9}$ of $3s.$ $9d.$ + $\cdot 4375$ of £1.

8. Find the volume of a partition 10 ft. long, 16 ft. high and 3 in. thick.

9. How many hours and minutes are there from 8.42 a.m. until 4.28 p.m.?

10. At $2d.$ a unit, how long can I allow a 25-watt lamp to burn so that it does not use more than $10d.$ worth of electricity?

ANSWERS

Ex. 22 A

1. 17 gal. 1 qt. 0 pt.
2. 131.
3. 10*d*.
4. $15\frac{3}{4}$ sq. in.
5. $\frac{1}{15}$.

6. 6.
7. 11*s*. 11*d*.
8. 27 cu. ft.
9. 7 hr. 35 min.
10. 50 hr.

Ex. 22 B

1. 13 yd. 2 ft. 6 in.
2. 62,000.
3. £11. 7*s*. 6*d*.
4. $30\frac{1}{4}$ sq. in.
5. $\frac{13}{15}$.

6. ·17.
7. 9*s*. 7*d*.
8. 40 cu. ft.
9. 7 hr. 46 min.
10. 200 hr.

NOTES AND MEMORANDA

More Progress Tests

Ex. 23 A

1. Take £3. 17s. 9½d. from £6. 7s. 11d.

2. 365 × 80.

3. How much shall I have to pay for 3½ gross at 8½d. each?

4. How many square inches are there in 1 square yard?

5. $\frac{18}{25} \div \frac{9}{20}$.

6. 5·076 + ·854.

7. If I start out with £·65 and spend 5s. 5½d., how much money shall I have left?

8. Find the volume of water which can be contained in a sink 6 ft. wide, 8 ft. long and 3 in. deep.

9. What is the cost of burning a 3500-watt fire for 8 hr. at 3d. a unit?

10. How many days were there from the beginning of July 1937 until the end of January 1938?

More Progress Tests (*continued*)

Ex. 23 B

1. From 17 tons 8 cwt. 2 qr. take 9 tons 15 cwt. 3 qr.

2. Change 67 in. to yards, etc.

3. At £3. 7s. 6d. for 4½ score, what is the price of 1?

4. How many square feet are there in an acre?

5. $\frac{16}{27} \times \frac{21}{32}$.

6. 7·835 − ·075.

7. I started out with $\frac{23}{40}$ of £1 and spent 4s. 6½d. How much money had I left?

8. What volume of water can be contained in a trough 5 ft. wide, 9 ft. long and 4 in. deep?

9. What is the cost of burning a 2500-watt fire for 12 hr. at 4d. a unit?

10. How many days were there from the beginning of April until the end of November?

ANSWERS

1. £2. 10s. 1½d.   2. 29,200.   3. £17. 17s. 0d.   4. 1296.   5. 1⅜.
6. 5·93.   7. 7s. 6¼d.   8. 12 cu. ft.   9. 7s.   10. 215.

Ex. 23 B

1. 7 tons 12 cwt. 3 qr.   2. 1 yd. 2 ft. 7 in.   3. 9d.   4. 43,560 sq. ft.   5. $\frac{7}{18}$.
6. 7·76.   7. 6s. 11½d.   8. 15 cu. ft.   9. 10s.   10. 244.

NOTES AND MEMORANDA

Percentages

A. Change to Percentages:

1. (a) $\frac{4}{5}$.   (b) $\frac{17}{100}$.
2. (a) $\frac{19}{20}$.   (b) $\frac{21}{25}$.
3. (a) ·9.   (b) ·875.
4. (a) ·065.   (b) ·3125.

Change to Vulgar Fractions:

5. (a) 60%.   (b) 85%.
6. (a) $37\frac{1}{2}$%.   (b) 48%.
7. (a) 74%.   (b) $57\frac{1}{2}$%.

Change to Decimals:

8. (a) 47%.   (b) 83%.
9. (a) $2\frac{1}{2}$%.   (b) 75%.
10. (a) $67\frac{1}{2}$%.   (b) $33\frac{1}{3}$%.

B. Change to Percentages:

1. (a) $\frac{3}{5}$.   (b) $\frac{23}{100}$.
2. (a) $\frac{17}{20}$.   (b) $\frac{37}{50}$.
3. (a) ·7.   (b) ·625.
4. (a) ·085.   (b) ·4375.

Change to Vulgar Fractions:

5. (a) 80%.   (b) 55%.
6. (a) $62\frac{1}{2}$%.   (b) 32%.
7. (a) 58%.   (b) $42\frac{1}{2}$%.

Change to Decimals:

8. (a) 39%.   (b) 91%.
9. (a) $7\frac{1}{2}$%.   (b) 25%.
10. (a) $42\frac{1}{2}$%.   (b) $66\frac{2}{3}$%.

Miscellaneous Exercises on Percentages

Ex. 25 A

1. 40% of 900.
2. $62\frac{1}{2}$% of 144.
3. 35% of 540.
4. 85% of £2.
5. $32\frac{1}{2}$% of £1.
6. $37\frac{1}{2}$% of 2s.
7. 70% of 1 ml. (in yd.).
8. 25% of 6 gal (in pt.).
9. From $57\frac{1}{2}$% of £1 take 70% of 10s.
10. Find the sum of 45% of £1 and $87\frac{1}{2}$% of £1.

Ex. 25 B

1. 60% of 800.
2. $87\frac{1}{2}$% of 128.
3. 15% of 740.
4. 65% of £3.
5. $27\frac{1}{2}$% of £1.
6. $12\frac{1}{2}$% of 5s.
7. 90% of 1 ton (in lb.).
8. 75% of 2 yd. (in in.).
9. From $72\frac{1}{2}$% of £1 take 90% of 10s.
10. Find the sum of 85% of £1 and $37\frac{1}{2}$% of £1.

ANSWERS

Ex. 24 A

1. (a) 80 %. (b) 17 %.
2. (a) 95 %. (b) 84 %.
3. (a) 90 %. (b) 87½ %.
4. (a) 6¼ %. (b) 31¼ %.
5. (a) $\frac{3}{8}$. (b) $\frac{7}{16}$.
6. (a) $\frac{1}{8}$. (b) $\frac{11}{13}$.
7. (a) $\frac{27}{50}$. (b) $\frac{23}{8}$.
8. (a) ·47. (b) ·83.
9. (a) ·025. (b) ·75.
10. (a) ·675. (b) ·$\dot{3}$.

Ex. 24 B

1. (a) 60 %. (b) 23 %.
2. (a) 85 %. (b) 74 %.
3. (a) 70 %. (b) 62½ %.
4. (a) 8¼ %. (b) 43¾ %.
5. (a) $\frac{4}{5}$. (b) $\frac{11}{20}$.
6. (a) $\frac{1}{8}$. (b) $\frac{8}{15}$.
7. (a) $\frac{23}{38}$. (b) $\frac{17}{40}$.
8. (a) ·39. (b) ·91.
9. (a) ·075. (b) ·25.
10. (a) ·425. (b) ·$\dot{6}$.

Ex. 25 A

1. 360. 2. 90. 3. 189. 4. £1. 14s. 0d. 5. 6s. 6d.
6. 9d. 7. 1232 yd. 8. 12 pt. 9. 4s. 6d. 10. £1. 6s. 6d.

Ex. 25 B

1. 480. 2. 112. 3. 111. 4. £1. 19s. 0d. 5. 5s. 6d.
6. 7½d. 7. 2016 lb. 8. 54 in. 9. 5s. 6d. 10. £1. 4s. 6d.

NOTES AND MEMORANDA

Profit and Loss, and Simple Interest

Ex. 26 A

1. If an article is bought for 10*s*. and sold for 13*s*., what is the gain per cent.?
2. Of 300 vegetable marrows 27 were bad. What was the percentage of good ones?
3. An article costing £4 is to be sold at a profit of $22\frac{1}{2}\%$. What will be the selling price?
4. A loss of 20 % was made by selling a book for 6*s*. What did the book cost?
5. A motor-car costing £150 was sold for £135. What was the percentage loss?

Find the Simple Interest on:

6. £100 for $2\frac{1}{2}$ years at 4 %.
7. £400 for 1 year at $3\frac{1}{2}\%$.
8. £300 for 4 years at $5\frac{1}{2}\%$.
9. £350 for 6 years at 3 %.
10. £25 for 8 years at $4\frac{1}{2}\%$.

Profit and Loss, and Simple Interest (*continued*)

Ex. 26 B

1. A book is bought for 5*s*. and sold for 7*s*. What is the percentage profit?
2. Only 88 % of 250 articles were fit for sale. What was the actual number of bad ones?
3. If an article costing £5 is to be sold at a profit of $17\frac{1}{2}\%$, what will be the selling price?
4. A loss of $12\frac{1}{2}\%$ was made by selling a knife for 3*s*. 6*d*. What did the knife cost?
5. A motor-cycle costing £50 was sold for £39. What was the loss per cent.?

Find the Simple Interest on:

6. £100 for $3\frac{1}{2}$ years at 6 %.
7. £500 for 1 year at $4\frac{1}{2}\%$.
8. £400 for $2\frac{1}{2}$ years at 5 %.
9. £250 for 5 years at 4 %.
10. £75 for 4 years at $2\frac{1}{2}\%$.

ANSWERS

1. 30 %.	2. 91 %.	3. £4. 18s. 0d.	4. 7s. 6d.	5. 10 %.
6. £10.	7. £14.	8. £66.	9. £63.	10. £9.

Ex. 26 B

1. 40 %.	2. 30.	3. £5. 17s. 6d.	4. 4s.	5. 22 %.
6. £21.	7. £22. 10s. 0d.	8. £50.	9. £50.	10. £7. 10s. 0d.

NOTES AND MEMORANDA

Commission and Discount

Ex. 27 A

1. Find the discount on £8 at $2\frac{1}{2}$%.

2. What percentage discount is one of 3s. 6d. in the £?

3. How much do I have to pay on a bill of 17s. 6d. if I am allowed a 15% discount?

4. At $17\frac{1}{2}$%, what is the discount on £5. 10s. 0d.?

5. Change 3s. 3d. in the £ to a percentage discount.

6. How much in the £ is a commission of 5%?

7. How much in the £ is a commission of $6\frac{1}{4}$%?

8. What percentage commission is 9d. in the £?

9. What does a commission of $7\frac{1}{2}$% on sales of £280 amount to?

10. How much is left after deducting a commission of 20% from sales amounting to £5?

Commission and Discount (*continued*)

Ex. 27 B

1. What is the discount on £6 at $7\frac{1}{2}$%?

2. Change a discount of 4s. 6d. in the £ to a percentage discount.

3. How much do I have to pay on a bill of £1. 7s. 6d. if I am allowed a 20% discount?

4. At $22\frac{1}{2}$%, what is the discount on £3. 10s. 0d.?

5. What percentage discount is 2s. 9d. in the £?

6. How much in the £ is a commission of 10%?

7. How much in the £ is a commission of $3\frac{3}{4}$%?

8. What percentage commission is 1s. 3d. in the £?

9. What does a commission of 15% on sales of £480 amount to?

10. What is left after deducting a commission of 15% from sales amounting to £12?

ANSWERS

Ex. 27 A

1. *4s.*   2. $17\frac{1}{2}$ %.   3. *14s. 10½d.*   4. *19s. 3d.*   5. $16\frac{1}{4}$ %.

6. *1s.*   7. *1s. 3d.*   8. $3\frac{3}{4}$ %.   9. £21.   10. £4.

Ex. 27 B

1. *9s.*   2. $22\frac{1}{2}$ %.   3. £1. 2s. 0d.   4. *15s. 9d.*   5. $13\frac{3}{4}$ %.

6. *2s.*   7. *9d.*   8. $6\frac{1}{4}$ %.   9. £72.   10. £10. 4s. 0d.

NOTES AND MEMORANDA

Sales' Reductions

Ex. 28 A. In the following questions you have the marked prices of articles and the amount of their reduction during sales. Find the actual amount you would have to pay for the articles during the sales:

1. Boys' shoes at 10s. per pair. Reduction of 1d. in 1s.
2. Girls' watches at 12s. 6d. each. Reduction of 1½d. in 1s.
3. Men's raincoats at £2. 15s. each. Reduction of 1s. in £1.
4. Women's hats at 17s. 6d. each. Reduction of 2s in £1.
5. Gold watches at £6. 10s. each. Reduction of 2s. 6d. in £1.
6. Tea services at £2. 10s. per set. Reduction of 15%.
7. Men's suits at £4. 15s. each. Reduction of 20%.
8. Boys' overcoats at £1. 18s. each. Reduction of 25%.
9. Gramophones at £4. 4s. each. Reduction of 12½%.
10. Wireless sets at £10 each. Reduction of 27½%.

Sales' Reductions *(continued)*

Ex. 28 B. In the following questions you have the marked prices of articles and the amount of their reduction during sales. Find the actual amount you would have to pay for the articles during the sales:

1. Men's hats at 15s. each. Reduction of 1d. in 1s.
2. Girls' gloves at 7s. 6d. per pair. Reduction of 1½d. in 1s.
3. Women's coats at £1. 15s. each. Reduction of 1s. in £1.
4. Boys' shoes at 12s. 6d. per pair. Reduction of 2s. in £1.
5. Men's suits at £5. 10s. each. Reduction of 2s. 6d. in £1.
6. Dinner services at £3. 10s. per set. Reduction of 15%.
7. Men's raincoats at £2. 15s. each. Reduction of 20%.
8. Girls' costumes at £2. 6s. each. Reduction of 25%.
9. Gramophones at £2. 16s. each. Reduction of 12½%.
10. Wireless sets at £12 each. Reduction of 22½%.

ANSWERS

Ex. 28 A

1. 9s. 2d.	2. 10s. 11½d.	3. £2. 12s. 3d.	4. 15s. 9d.	5. £5. 13s. 9d.
6. £2. 2s. 6d.	7. £3. 16s. 0d.	8. £1. 8s. 6d.	9. £3. 13s. 6d.	10. £7. 5s. 0d.

Ex. 28 B

1. 13s. 9d.	2. 6s. 6¼d.	3. £1. 13s. 3d.	4. 11s. 3d.	5. £4. 16s. 3d.
6. £2. 19s. 6d.	7. £2. 4s. 0d.	8. £1. 14s. 6d.	9. £2. 9s. 0d.	10. £9. 6s. 0d.

NOTES AND MEMORANDA

Proportion

1. At 7s. 6d. per 2 dozen, what is the price of 1?
2. What will 1 cost at £16. 5s. per 2½ score?
3. If 9 articles cost 3s. 9d., what will be the cost of 1¼ dozen?
4. 3½ lb. cost 10s. 6d. What is the cost of 5½ lb.?

The cost carrying 12 tons 5 miles is £4. For the same money find:

5. How far 5 tons should be carried.
6.   ,,   6   ,,   ,,
7.   ,,   4   ,,   ,,
8.   ,,   20   ,,   ,,
9. In 2 weeks 5 men earn £30. What will 2 men earn in 3 weeks?
10. 100 ft. are built by 4 men in 5 days. At the same rate how many feet can be built by 20 men in 3 days?

Proportion (*continued*)

1. At 12s. 6d. per 2 dozen, what is the price of 1?
2. What will 1 cost at £23. 15s. per 2½ score?
3. If 11 articles cost 2s. 9d., what will 1¾ dozen cost?
4. 4½ lb. cost 11s. 3d. What will 2½ lb. cost?

The cost of carrying 10 tons 9 miles is £5. For the same money find:

5. How far 2 tons should be carried.
6.   ,,   3   ,,   ,,
7.   ,,   5   ,,   ,,
8.   ,,   20   ,,   ,,
9. In 3 weeks 4 men earn £48. What will 3 men earn in 1 week?
10. 5 people travel 20 miles for £1. How many people can travel 5 miles for £3?

ANSWERS

Ex. 29 A

1. 3¼d. 2. 6s. 6d. 3. 6s. 3d. 4. 16s. 6d. 5. 12 ml.

6. 10 ml. 7. 15 ml. 8. 3 ml. 9. £18. 10. 300 ft.

Ex. 29 B

1. 6¼d. 2. 9s. 6d. 3. 5s. 3d. 4. 6s. 3d. 5. 45 ml.

6. 30 ml. 7. 18 ml. 8. 4½ ml. 9. £12. 10. 60.

NOTES AND MEMORANDA

More Examples on Proportion

Ex. 30 A

1. If 12 articles cost 13s. 9d., what do 7 cost?
2. 7 articles cost 4s. 11½d. What do 12 cost?
3. At 5 for 6d., what is the cost of 10 dozen?
4. What is the cost of 54 at 17s. 6d. for 7 dozen?
5. How long will rations for 9 men for 4 days last 12 men?
6. If 6 men do a job in 3 days, how long should it take 9 men?
7. At 30 m.p.h. a journey is completed in 4 hr. How long would it have taken at 40 m.p.h.?
8. 5 tons are carried 6 miles for £1. How far should 15 tons be carried for the same money?
9. 8 people travel 20 miles for £1. How far can 16 people travel for £4?
10. In 10 weeks 3 men earned £90. At this rate what will 2 men earn in 4 weeks?

More Examples on Proportion (*continued*)

Ex. 30 B

1. If 12 articles cost 17s. 3d., what do 5 cost?
2. 5 articles cost 4s. 9½d. What do 12 cost?
3. At 8 for 6d., what is the cost of 1 gross?
4. What is the cost of 66 at 17s. 6d. for 5 dozen?
5. How long will rations for 8 men for 6 days last 24 men?
6. If 4 men can do a job in 6 days, how long should it take 12 men?
7. At 20 m.p.h. a journey is completed in 3 hr. How long would it have taken at 30 m.p.h.?
8. 4 tons are carried 18 miles for £5. How far should 24 tons be carried for the same money?
9. 10 people can travel 16 miles for £1. How far can 40 people travel for £5?
10. In 3 weeks 8 men earned £96. At this rate what will 4 men earn in 4 weeks?

ANSWERS

Ex. 30 A

1. 8*s.* 0¼*d.*	2. 8*s.* 6*d.*	3. 12*s.*	4. 11*s.* 3*d.*	5. 3 dy.
6. 2 dy.	7. 3 hr.	8. 2 ml.	9. 40 ml.	10. £24.

Ex. 30 B

1. 7*s.* 2¼*d.*	2. 11*s.* 6*d.*	3. 9*s.*	4. 19*s.* 3*d.*	5. 2 dy.
6. 2 dy.	7. 2 hr.	8. 3 ml.	9. 20 ml.	10. £64

NOTES AND MEMORANDA

Arithmetic in Daily Life

Ex. 31 A

1. What change is received from £1 after paying a bill of 13s. 3½d.?
2. Find the cost of 3¾ lb. at 1s. 8d. per lb.
3. What volume of water (in cubic in.) can be contained in a pipe of 1 in. inside radius and length of 2 ft. 6 in.? ($\pi = 3 \cdot 14$.)
4. At 1s. 2d. per hr., what are the wages of a man working 42 hr.?
5. What is the cost of burning ten 60-watt lamps for 5 hr. each at 2¾d. a unit?
6. Find the cost of 4 gal. 3 qt. 1 pt. at 4d. per gal.
7. What is the total annual insurance premium on property valued at £650 at 2s. 4d. per cent.?
8. What percentage commission is 1s. 9d. in the £?
9. Find the sale price of a £4 coat reduced 17½%.
10. What is the cost of 2½ lb. at 1s. 10d.?

Arithmetic in Daily Life (*continued*)

Ex. 31 B

1. Find the change received from £1 after paying a bill of 12s. 7½d.
2. What is the cost of 2¾ lb. at 2s. 4d. per lb.?
3. What volume of water (in cubic ft.) can be contained in a pipe of 2 ft. inside radius and length 10 ft. ($\pi = 3 \cdot 14$.)
4. Find the total time from 8.15 a.m. to 12.30 p.m. and from 1.45 p.m. to 5.30 p.m.
5. What is the cost of burning a 3500-watt fire 8 hr. at 2¾d. a unit?
6. What is the hire purchase price of a gramophone at 15s. 0d. down and 11 weekly instalments of 2s. 6d.?
7. Find the total annual insurance premium on property valued at £850 at 2s. 2d. per cent.
8. How much is a discount of 32½% in the £?
9. A £5 article is reduced 7s. 6d. What is the percentage reduction?
10. What is the cost of 3½ lb. at 1s. 4d. per lb.?

ANSWERS

Ex. 31 A

1. 6s. 8¼d.	2. 6s. 3d.	3. 94·2 cu. in.	4. £2. 9s. 0d.	5. 8¼d.
6. 1s. 7½d.	7. 15s. 2d.	8. 8¾ %.	9. £3. 6s. 0d.	10. 4s. 7d.

Ex. 31 B

1. 7s. 4½d.	2. 6s. 5d.	3. 125·6 cu. ft.	4. 8 hr.	5. 6s. 5d.
6. £2. 2s. 6d.	7. 18s. 5d.	8. 6s. 6d.	9. 7½ %.	10. 4s. 8d.

NOTES AND MEMORANDA

Algebra

Ex. 32 A. Simplify:

1. $23a + 16a + 9a$.
2. $73x - 47x$.
3. $4m \times 17m$.
4. $144y^2 \div 8y$.
5. $16b - (4b + 3b)$.
6. $15a + 2(3a - 4)$.
7. $11y = 132$. Find the value of y.
8. $\dfrac{3x}{4} = 6$. Find the value of x.
9. What is the area of a rectangle $5y$ ft. by $13x$ ft.?
10. What is the width of a rectangle $8s$ ft. wide with an area of $56st$ sq. ft.?

B. Simplify:

1. $7b + 32b + 15b$.
2. $82y - 58y$.
3. $5n \times 18n$.
4. $133a^2 \div 7a$.
5. $15x - (6x - 2x)$.
6. $13a + 4(2a + 3)$.
7. $14x = 42$. Find the value of x.
8. $\dfrac{4b}{5} = 8$. Find the value of b.
9. What is the perimeter of a rectangle $2a$ ft. by $3b$ ft.?
10. What is the length of a rectangle $6a$ ft. wide with an area of $78ab$ sq. ft.?

Algebra (*continued*)

Ex. 33 A

1. $31y + 17y + 28y$.
2. $124b - 87b$.
3. $7r \times 9r \times 11r$.
4. $96x^2y \div 4x$.
5. $7a - 3(2a + 4)$.
6. $8x + 5(x - 4)$.
7. $14m - 3 = 25$. Find the value of m.
8. $3x^2 = 75$. Find the value of x.
9. $2a^2 - 3 = 29$. Find the value of a.
10. If I walk $8x$ ml. in 4 hr., what is my speed in m.p.h.?

Ex. 33 B

1. $16x + 27x + 33x$.
2. $135a - 78a$.
3. $6p \times 8p \times 12p$.
4. $95ab^2 \div 5b$.
5. $5b + 4(6b - 2)$.
6. $9y - 2(3y + 2)$.
7. $15n - 5 = 40$. Find the value of n.
8. $4y^2 = 64$. Find the value of y.
9. $3b^2 - 4 = 23$. Find the value of b.
10. If I sell $5p$ articles out of $7m$ articles, how many are left?

ANSWERS

Ex. 32 A

1. $48a$.
2. $26x$.
3. $68m^2$.
4. $18y$.
5. $9b$.
6. $21a - 8$.
7. $y = 12$.
8. $x = 8$.
9. $65xy$ sq. ft.
10. $7t$ ft.

Ex. 32 B

1. $54b$.
2. $24y$.
3. $90n^2$.
4. $19a$.
5. $11x$.
6. $21a + 12$.
7. $x = 3$.
8. $b = 10$.
9. $(4a + 6b)$ ft.
10. $13b$ ft.

Ex. 33 A

1. $76y$.
2. $37b$.
3. $693r^2$.
4. $24xy$.
5. $a - 12$.
6. $13x - 20$.
7. $m = 2$.
8. $x = 5$.
9. $a = 4$.
10. $2x$ m.p.h.

Ex. 33 B

1. $76x$.
2. $57a$.
3. $576p^2$.
4. $19ab$.
5. $29b - 8$.
6. $3y - 4$.
7. $n = 3$.
8. $y = 4$.
9. $b = 3$.
10. $7m - 5p$.

NOTES AND MEMORANDA

Revision Tests in Arithmetic

Ex. 34 A

1. £3. 19s. 2½d. + £2. 16s. 4¼d. + £4. 8s. 7½d.

2. 67 × 102.

3. Find the cost of 3¼ dozen articles at 10½d. each.

4. ⅔ of £1. 17s. 4d.

5. 3·067 + ·973.

6. What is left after taking from 240, 37½ % of it?

7. Find the percentage profit made on an article bought for 6s. and sold for 6s. 9d.

8. At 15 %, what is the discount on a bill of £12. 5s.?

9. What is the sale price of a £5. 10s. article which has been reduced 20 %?

10. If 8 books cost 14s., what will 1½ dozen cost?

Revision Tests in Arithmetic (*continued*)

Ex. 34 B

1. 22 tons 16 cwt. 3 qr. ÷ 9.

2. How many threepences are there in £1. 17s. 9d.?

3. What is the cost of 3½ gross articles at 9¼d. each?

4. ·025 of £3. 10s.

5. 4/7 − 3/5.

6. What percentage of 360 is 9?

7. Find the selling price of an article bought for 10s. and sold at a profit of 7½ %.

8. What percentage commission is 2s. 3d. in the £?

9. What is the sale price of a £13. 10s. article which has been reduced 12½ %?

10. If 5 knives cost 12s. 1d., what will 1½ dozen cost?

ANSWERS

Ex. 34A

1. £11. 4s. 2¼d.
2. 6834.
3. £1. 14s. 1½d.
4. 10s. 8d.
5. 4·04.
6. 150.
7. 12½ %.
8. £1. 16s. 9d.
9. £4. 8s. 0d.
10. £1. 11s. 6d.

Ex. 34B

1. 2 tons 10 cwt. 3 qr.
2. 151.
3. £19. 8s. 6d.
4. 1s. 9d.
5. $\frac{6}{35}$.
6. 2¼ %.
7. 10s. 9d.
8. 11¼ %.
9. £11. 16s. 3d.
10. £2. 3s. 6d.

NOTES AND MEMORANDA

Revision Tests in Mensuration

Ex. 35 A

1. What is the perimeter of a rectangle $11\frac{3}{4}$ in. by $7\frac{1}{2}$ in.?
2. Find the area of a square with a side of $9\frac{1}{2}$ in.
3. A triangle has a base of 5 in. and a height of 9 in. What is its area?
4. A circle has a radius of $4\frac{1}{2}$ in. What is its circumference? ($\pi = 3\cdot14$.)
5. Find the area of a circle with a 10 in. diameter.
6. Change $2\frac{1}{2}$ sq. ft. to square inches.
7. $\sqrt{4900} - \sqrt{121}$.
8. How many cubic feet are there in 12 cubic yd.?
9. What is the volume of a cylinder 1 ft. 8 in. long and with a radius of 2 in.? ($\pi = 3\cdot14$.)
10. A triangle with a base of 8 in. has an area of 20 sq. in. What is its height?

Revision Tests in Mensuration (*continued*)

Ex. 35 B

1. What is the perimeter of a rectangle $9\frac{1}{2}$ in. by $10\frac{1}{4}$ in.?
2. Find the area of a rectangle $6\frac{1}{2}$ in. by $5\frac{1}{2}$ in.
3. The base of a triangle is 8 in. and its length 1 ft. 4 in. What is its area?
4. A circle has a circumference of $6\cdot28$ in. What is its radius? ($\pi = 3\cdot14$.)
5. What is the area of a circle with a 20 in. radius?
6. How many square inches are there in 6 sq. ft.?
7. $90^2 + 40^2$.
8. Change 2420 sq. ft. to square yards, etc.
9. What is the volume of a cylinder 10 in. long with a radius of 5 in.? ($\pi = 3\cdot14$.)
10. A rectangular field has an area of 1 acre. One side is 40 yd., what is the length of the other?

ANSWERS

Ex. 35 A

1. 3 ft. 2½ in.
2. 90¼ sq. in.
3. 22½ sq. in.
4. 28·26 in.
5. 78·5 sq. in.
6. 360 sq. in.
7. 59.
8. 324 cu. ft.
9. 251·2 cu. in.
10. 5 in.

Ex. 35 B

1. 3 ft. 3½ in.
2. 35¾ sq. in.
3. 64 sq. in.
4. 1 in.
5. 1256 sq. in.
6. 864.
7. 9700.
8. 268 sq. yd. 8 sq. ft.
9. 785 cu. in.
10. 121 yd.

NOTES AND MEMORANDA

Revision Tests in Algebra

Ex. 36 A

1. $16a - 18a + 23a$.
2. $12x \times 8xy$.
3. $19m^2 \times 6m$.
4. $196ab^2 \div 7ab$.
5. $54x^3 \div 27x^2$.
6. $24a - 3(7a + 4)$.
7. $16x + 5(3x + 2)$.
8. $7x + 9 = 58$. Find the value of x.
9. $\dfrac{4y}{7} = 8$. Find the value of y.
10. What is the perimeter of a rectangle $5x$ ft. by $3y$ ft.?

Ex. 36 B

1. $15b - 17b + 27b$.
2. $11y \times 6xy$.
3. $17r^2 \times 8r$.
4. $184x^2y \div 8xy$.
5. $64a^3 \div 16a^2$.
6. $32b - 4(8b - 4)$.
7. $18y + 7(2y - 4)$.
8. $11a + 8 = 74$. Find the value of a.
9. $\dfrac{5b}{6} = 15$. Find the value of b.
10. What is the area of a rectangle $6a$ ft. by $7a$ ft.?

Revision Tests in Algebra (*continued*)

Ex. 37 A

1. $27x - 30y + 42y$.
2. $8x^2y \times 9xy$.
3. $16st \times 8r$.
4. $72abc \div 18c$.
5. $144xy \div 24y$.
6. $9a - (3a + 6)$.
7. $8y + 3x - 2(4y + x)$.
8. $11m - 10 = 89$. Find the value of m.
9. $\dfrac{3a - 3}{5} = 3$. Find the value of a.
10. A rectangle $18xy^2$ sq. ft. in area has a width of $6xy$ ft. What is its length?

Ex. 37 B

1. $29x - 35x + 51x$.
2. $12ab^2 \times 7ab$.
3. $17ef \times 6g$.
4. $84xyz \div 14z$.
5. $132ab \div 22a$.
6. $12b - (7b - 6)$.
7. $9a - 4b - 3(3a - 6b)$.
8. $12n - 11 = 133$. Find the value of n.
9. $\dfrac{2x - 4}{3} = 6$. Find the value of x.
10. A rectangle is $27ab^2$ sq. yd. in area. Its length is $9ab$ ft. What is its width?

ANSWERS

Ex. 36 A

1. $21a$. 2. $96x^2y$. 3. $114m^3$. 4. $28b$. 5. $2x$.

6. $3a-12$. 7. $31x+10$. 8. $x=7$. 9. $y=14$. 10. $(10x+6y)$ ft.

Ex. 36 B

1. $25b$. 2. $66xy^2$. 3. $136r^3$. 4. $23x$. 5. $4a$.

6. 16. 7. $32y-28$. 8. $a=6$. 9. $b=18$. 10. $42a^2$ sq. ft.

Ex. 37 A

1. $39y$. 2. $72x^3y^3$. 3. $128rst$. 4. $4ab$. 5. $6x$.

6. $6a-6$. 7. x. 8. $m=9$. 9. $a=6$. 10. $3y$ ft.

Ex. 37 B

1. $45x$. 2. $84a^2b^3$. 3. $102\ efg$. 4. $6xy$. 5. $6b$.

6. $5b+6$. 7. $14b$. 8. $n=12$. 9. $x=11$. 10. $3b$ ft.

NOTES AND MEMORANDA

Progress Tests

Ex. 38 A

1. 5 gal. 2 qt. 1 pt. + 3 gal. 3 qt. 0 pt. + 7 gal. 1 qt. 1 pt.

2. Change £29. 16s. 8d. to three-and-fourpences.

3. Find the cost of 1 at £24. 6s. 0d. per 4½ gross.

4. ·2 of 2 tons 6 cwt. 1 qr.

5. $\frac{3}{8}$ of 2s. 8d. + ·25 of 6s.

6. What is the Simple Interest on £450 for 2 years at 3½ %?

7. How many hours and minutes are there from 6.45 a.m. to 5.20 p.m.?

8. What is the depth of a tank with a volume of 60 cubic ft., and an inside length and breadth of 6 ft. and 4 ft.?

9. What is the value of a 7½ % commission on sales of £35. 10s. 0d.?

10. 7 men build 200 ft. in 6 days. How much can be built by 21 men in 3 days?

Progress Tests (*continued*)

Ex. 38 B

1. 23 yd. 1 ft. 7 in. − 18 yd. 2 ft. 10 in.

2. 54 × 98.

3. What is the cost of 5½ score at 5s. 6d. each?

4. $\frac{1}{8}$ of 3 yd. 2 ft. 9 in.

5. $\frac{4}{5}$ of 6s. 3d. + ·625 of 2s.

6. If I buy a motor-car for £160 and sell it at a loss of 37½ %, what do I get for it?

7. What do I have to pay for a £6. 16s. 0d. article which is being sold at a reduction of 2s. 6d. in the £?

8. Find the area of a triangle with base $2x$ in. and height $6y$ in.

9. What do I have to pay on a bill of £4 which is subject to a discount of 17½ %?

10. Working at the same rate as 12 men who complete a job in 16 days, how long will it take 8 men to do a similar job?

[95]

ANSWERS

Ex. 38 A

1. 16 gal. 3 qt. 0 pt.
2. 179.
3. 9*d.*
4. 9 cwt. 1 qr.
5. 2*s.* 6*d.*
6. £31. 10*s.* 0*d.*
7. 10 hr. 35 min.
8. 2½ ft.
9. £2. 13*s.* 3*d.*
10. 300 ft.

Ex. 38 B

1. 4 yd. 1 ft. 9 in.
2. 5292.
3. £30. 5*s.* 0*d.*
4. 1 ft. 11½ in.
5. 6*s.* 3*d.*
6. £100.
7. £5. 19*s.* 0*d.*
8. 6*xy* sq. in.
9. £3. 6*s.* 0*d.*
10. 24 dy.

NOTES AND MEMORANDA

Final Tests

Ex. 39 A

1. £25. 15s. 3d. − £17. 18s. 7½d.

2. 776 × 125.

3. Find the cost of 2¾ dozen articles at 2½d. each.

4. ³⁄₇ of 15s. 9d.

5. If I bought an article for £40 and sold it at a profit of 15 %, what was my selling price?

6. What is the hire purchase price of an article costing £1. 10s. 0d. down and 8 monthly instalments of 17s. 6d.?

7. Find the area of a circle with a diameter of 6x in.

8. How much do I have to pay on a bill of £8 which is subject to a discount of 22½ %?

9. If 15 articles cost £3. 15s. 0d., what will be the cost of 2 dozen?

10. What is earned by a man who works 49 hr. at 1s. 6d. per hr.?

Final Tests (*continued*)

Ex. 39 B

1. £3. 19s. 6½d. + £2. 6s. 4¼d. + £18. 3s. 9½d.

2. 9625 ÷ 125.

3. What is the cost of 1 at £30. 12s. 6d. per 2½ score?

4. ·075 of £5. 10s. 0d.

5. Find the Simple Interest on £550 for 1½ years at 6 %.

6. How much is earned by a man who works 52 hr. at 1s. 3d. per hr.?

7. What is the volume (in cu. in.) of a cylinder 4 ft. 2 in. long with a diameter of 20 in.?

8. What is the value of a 2½ % commission on sales of £17. 5s. 0d.?

9. At 10 m.p.h. a cyclist completes his journey in 3 hr. How long would it have taken him if he had averaged 15 m.p.h.?

10. How many hours does a man work in 5 days working from 7.30 a.m. till 12.15 p.m. and from 1.30 p.m. till 5.45 p.m. daily?

ANSWERS

Ex. 39 A

1. £7. 16s. 7½d.   2. 97,000.   3. 6s. 10½d.   4. 6s. 9d.   5. £46.

6. £8. 10s. 0d.   7. 28·26x^2 sq. in.   8. £6. 4s. 0d.   9. £6.   10. £3. 13s. 6d.

Ex. 39 B

1. £24. 9s. 8¼d.   2. 77.   3. 12s. 3d.   4. 8s. 3d.   5. £49. 10s. 0d.

6. £3. 5s. 0d.   7. 15,700 cu. in.   8. 8s. 7½d.   9. 2 hr.   10. 45 hr.

NOTES AND MEMORANDA

Printed in the United States
By Bookmasters